EARTH SHELTERED HOMES

PLANS AND DESIGNS

UNDERGROUND SPACE CENTER
UNIVERSITY OF MINNESOTA

EARTH SHELTERED HOMES

PLANS AND DESIGNS

UNDERGROUND SPACE CENTER
UNIVERSITY OF MINNESOTA

DONNA AHRENS

TOM ELLISON
Principal, Ellison Design and Construction
Minneapolis, Minnesota

RAY STERLING

VNR VAN NOSTRAND REINHOLD COMPANY
NEW YORK CINCINNATI TORONTO LONDON MELBOURNE

Printed in the United States of America

Published by Van Nostrand Reinhold Company
135 West 50th Street
New York, NY 10020

Van Nostrand Reinhold Limited
1410 Birchmount Road
Scarborough, Ontario M1P 2E7, Canada

Van Nostrand Reinhold Australia Pty. Ltd.
17 Queen Street
Mitcham, Victoria 3132, Australia

Van Nostrand Reinhold Company Limited
Molly Millars Lane
Wokingham, Berkshire, England

16 15 14 13 12 11 10 9 8 7 6 5 4 3 2

Library of Congress Cataloging in Publication Data
Minnesota. University. Underground Space Center.
      Earth sheltered homes.

      1. Earth sheltered houses—Design and construction.
I. Ahrens, Donna. II. Title.
TH4819.E27M56 1981   728   80-28327
ISBN 0-442-28675-9
ISBN 0-442-28676-7 (pbk.)

Preface

The Underground Space Center decided to prepare this second book on earth sheltered homes for a number of reasons. Since the publication in 1978 of *Earth Sheltered Housing Design*, which featured earth sheltered houses built prior to 1977, both the public and the building industry have expressed a strong desire to examine a greater variety of earth sheltered designs. *Earth Sheltered Homes* provides the opportunity to do just that.

In 1976, when research for *Earth Sheltered Housing Design* began, earth sheltered building technology was unfamiliar to all but a handful of architects and builders. Today—thanks to a combination of dramatic growth in the earth sheltered construction industry and extensive media coverage of this innovative type of design—earth sheltering is, if not yet a household phrase, a concept familiar to a great many people. Earth sheltered residences have been built in virtually every region of the United States, displaying a wide range of styles, sizes, and construction techniques.

The publicity that typically has followed the completion of such homes, however, has proven to be a double-edged sword: while it has piqued the curiosity of the general public regarding underground homes, it has also created an increasing reluctance on the part of home owners to open their doors continually to ever greater numbers of interested passersby. This dearth of opportunities to visit one or more well-designed earth sheltered homes is unfortunate, since such firsthand inspections often serve as a major turning point in people's perceptions and assessments of earth sheltered housing.

While pictures, plans, and words cannot, of course, duplicate the experience of visiting an earth sheltered dwelling, this book aims to give the public an armchair tour of a variety of earth sheltered homes from different parts of the United States and Europe. For the practicing or potential designer, the visual images, together with the plans and construction information included in these pages, should provide a ready reference of design ideas and solutions.

RAY STERLING
Director, The Underground Space Center

During the course of writing and designing this book, the authors received generous support from a number of individuals.

Earth Sheltered Homes could not have been completed without extensive assistance from the architects who designed the twenty-three earth sheltered houses featured in the book. We gratefully acknowledge the time, effort, and money they spent in providing us with the drawings, photographs, and written information we requested.

Rick LaMuro was a most valuable design assistant, devoting long hours to preparing detailed drawings and layouts. His patience and good nature throughout the design process were much appreciated.

Several other Underground Space Center staff members who contributed significantly to preparation of the book are Arlene Bennett (typing—and retyping!), Nancy Larson (initial information gathering), and Suzanne Swain (keylining).

Special thanks go to John Carmody for his suggestions regarding the conceptualization and execution of the project, and to Leslie Roberts for critically reading the overview section of the manuscript.

We also wish to thank the following individuals and organizations for granting permission to use their photographs and drawings in the *Overview:* Malcolm Wells—sketches on pages 8, 14, and 17 (from *Underground Designs* and *Underground Plans Book I*); Ezra Stoller, © ESTO—pages 13 and 14, Geier house (Philip Johnson, architect); and the Minnesota Historical Society, H. D. Ayer, photographer—page 15 (Midwestern sod house).

Acknowledgments

Contents

Part 1

Overview of Earth Sheltered Housing — 8

Part 2

Minnesota Housing Finance Agency Demonstration Houses
Introduction — 21

Camden State Park House — 22
Willmar House — 26
Waseca House — 30
Burnsville House — 34
Seward Town Houses — 38
Wild River State Park House — 42
Whitewater State Park House — 46

Part 3

Sixteen Houses from the U.S. and Europe
Introduction — 53

Suncave — 54
Remington House — 58
Clark House — 62
Hadley House — 66
Feuille House — 70
Sticks & Stones House — 74
Earthtech 5 and 6 — 78
SunEarth House — 84
Boothe House — 88
Wells House/Office — 92
Architerra Houses — 96
Terra-Dome House — 102
Demuyt House — 106
Wheeler House — 110
Moreland House — 114
Topic House — 118

Appendix — 123
Index — 125

Overview of Earth Sheltered Housing Part 1

In 1974 a conference sponsored by the University of Minnesota on earth sheltered construction was attended by a total of twenty people. Three years later a similar conference drew about four hundred people and spawned the Underground Space Center, a research and information organization based at the University of Minnesota.

By 1980 two April conferences on earth sheltering—one in Minnesota and another in Oklahoma—together registered more than one thousand participants, and an earth sheltered housing exhibition of products and services at the Minnesota conference attracted a crowd of nearly two thousand.

Evidence of the fast-rising popularity of this architectural style, which incorporates earth into the house design, is the increasing number of books dealing with the subject. After the Underground Space Center published its *Earth Sheltered Housing Design*, several magazines that deal solely with aspects of underground construction and planning, such as *Earth Sheltered Digest* and *Underground Space*, were established, and numerous magazine and newspaper articles concerned with earth sheltered design and construction have appeared. Thousands of people continue to flock to public open houses to see demonstration earth sheltered homes; a growing number of conferences, workshops, and courses deal with a wide range of earth sheltering applications; and at least a dozen universities now have programs and centers involved in research and development as they relate to various aspects of underground space use.

The most concrete evidence—literally and figuratively—of the growing acceptance of earth sheltered housing is the increasing numbers of such homes being built today. The American Institute of Architects estimates that by 1978 only thirty to forty earth sheltered houses had been built; by mid-1980 that number had grown to three thousand to five thousand. And if inquiries to the Underground Space Center are a reliable indicator, thousands more such homes are in the planning stages. To date, the Underground Space Center has received over thirty-five thousand requests from all over the world for information about earth sheltered and underground design and construction—and this enthusiasm shows no sign of abating. Furthermore, entire earth sheltered communities are in the planning or construction stages in Wisconsin and Washington, and large-scale housing developments are underway in Alabama, California, and Minnesota.

Given an established and increasing demand for earth sheltered housing, an industry has sprung up to fill that need. In the last four years, a number of contractors and developers have begun specializing in earth sheltered homes; about half a dozen of these report being involved in the planning and/or construction of one hundred to two hundred homes each. These builders generally offer packaged designs, better cost control, and an ability to improve their techniques from project to project.

Why Earth Sheltering?

Ask any expert in the field of earth sheltered construction about the reasons for this upsurge of interest in underground living, and you will probably get a two-word answer: *energy costs*. The rising costs of fossil fuels have without doubt given impetus to the recent growth of the earth sheltered housing industry. Although reliable energy performance data are limited, experts have estimated that earth sheltered houses use 25 to 80 percent of the energy required by pre-energy-crisis houses, which constitute the majority of our present housing stock. Some earth sheltered home owners have reported winter energy costs as low as $1.20 (the cost of cutting 2½ cords of wood); however, most northern-climate subterranean homes require some form of electrical backup heating.

Obviously, earth sheltering is not the only means of saving energy in housing—homes that appropriately use superinsulation, active solar, and/or passive solar can achieve energy performances similar to those of earth sheltered residences. Earth sheltered homes do have additional advantages, however. For example, they are generally much quieter than conventional, aboveground houses because the earth surrounding them "dampens" noise from the outside. The masonry/concrete structure (concrete is still the most commonly used structural material) is rot- and vermin-proof and usually more fire-resistant than materials used in above-grade houses. Moreover, because these structures are below grade level, natural disasters such as tornadoes and severe storms have less effect on them.

Earth sheltering makes good environmental sense too. By building into a hillside or below the earth's surface, an attractive landscape or view can be preserved while allowing access to natural light. Furthermore, sites that may be undesirable for conventional homes—for example, due to noise or traffic patterns—may be successfully adapted for residential use through earth sheltering. The Seward town houses (see pages 38-41) provide an excellent illustration both of surface preservation and effective use of a site considered "undesirable" by conventional building standards.

Another environmental plus associated with earth sheltering is the concept of working with nature as part of the design plan. In fact, working with the site through design and landscaping so the house will blend into the surrounding environment is part and parcel of the overall concept of earth sheltering. Thus, the lines and forms of earth sheltered houses tend to complement and duplicate forms found in nature.

Earth Sheltering—How and Why It Works

People unfamiliar with the basic concepts of earth sheltering often think that the excellent thermal performance associated with underground houses results from insulating qualities of the earth around them. In fact, although the large amounts of earth that usually cover three or more sides and the roofs of most earth sheltered houses do have an insulating effect, many feet of earth would be required to equal the insulating properties of just a few inches of rigid insulation. Rather, the energy-saving potential of earth sheltered homes is based on several of their physical characteristics.

First, earth sheltered houses lose less heat through the walls and roof of the building than do conventional aboveground structures. Conventional houses lose heat to the colder outside air in winter and gain heat from the hot outside air in summer. In contrast, the earth surrounding an underground structure works as a temperature moderator, reducing summer heat gain and winter heat loss. The relatively stable temperature of the soil surrounding an earth sheltered house means that in summer the house loses heat to the cool earth rather than gaining heat from the surrounding air, and in winter the relatively warm soil offers a much better temperature environment than the subzero air temperatures.

Burnsville house, Minneapolis, Minnesota.

Wild River State Park house, Minnesota.

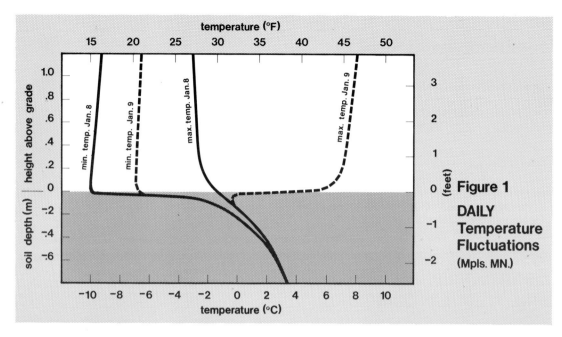

Figure 1

DAILY Temperature Fluctuations

(Mpls. MN.)

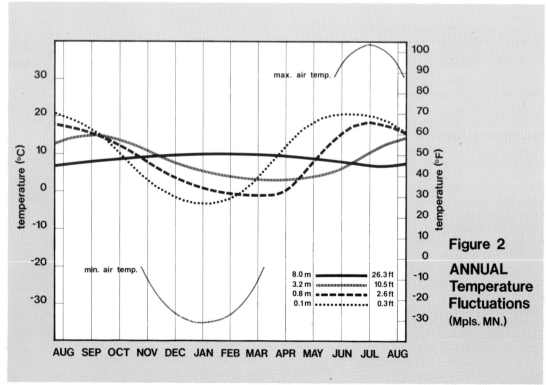

Figure 2

ANNUAL Temperature Fluctuations

(Mpls. MN.)

8.0 m	26.3 ft
3.2 m	10.5 ft
0.8 m	2.6 ft
0.1 m	0.3 ft

This concept is illustrated by Figures 1 and 2, indicating the daily and yearly soil temperature fluctuations at various depths. Figure 1 shows that daily fluctuations are virtually eliminated even at a depth of 8 inches (20 cm) of soil. At greater depths, soil temperatures respond only to seasonal changes, and the temperature change occurs after considerable delay.

Figure 2 indicates the seasonal temperature fluctuations at different depths for the Minneapolis-St.Paul area. Here, where outside air temperature swings as much as 130°F (72°C)—from -30° to 100°F (-34° to 38°C)—annually, the temperature of the soil 17 to 26 feet (5 to 8 m) below the surface is virtually constant. Ten feet (3 m) below the surface, the soil temperature varies from only 40° to 60°F (4° to 16°C); and even immediately below the surface, the annual soil temperature range is only 40°F (22°C).

The slowness with which soil temperatures change creates a thermal flywheel effect that contributes significantly to the energy efficiency of earth sheltered dwellings. In Minnesota, soil 10 feet (3 m) below the ground reaches its coldest temperature not in the dead of winter but in early spring, just as air temperatures begin to warm up. By the same principle, this soil is warmest around November, when outside temperatures begin to drop. Hence, the periods when energy derived from fossil fuels is likely to be necessary are shorter than is the case for most conventional houses.

Another energy-saving feature characteristic of earth sheltered structures, as distinguished from aboveground structures, is the lower heat loss due to infiltration. A conventional above-grade house loses a certain amount of its heat through cracks around windows and doors and generally throughout the structure—a process that is accelerated when the wind blows. With proper siting, the earth can protect an earth sheltered house from the wind, thus reducing general infiltration—and heating bills—considerably.

A final important characteristic of earth sheltered houses is the high thermal mass of the structure and the surrounding earth, which contributes to the heat storage capacity of a building. The thermal mass of a structure is a function of the density and quantity of the building materials in combination with the ability of those materials to store heat. A house with a larger thermal mass, especially one with a concrete shell, can absorb heat from the air or from direct solar radiation. This heat can then be released back into the space during the night, when there is a net heat loss. In an earth sheltered house, which has a high thermal mass, this process can be slow enough to "carry" the house for several hours without heat from an additional source. A conventional home, on the other hand, can store very little excess heat gain and loses whatever heat it has relatively rapidly when a heat source is interrupted.

Temperature data collected over a four-day period from an earth sheltered house in Rolla, Missouri, illustrate this effect. Figure 3 shows that the inside temperature remained relatively stable in the absence of any internal heating while the outside temperature dropped to -10°F (-23°C). Monitoring data for the Topic house (pages 118-121) bear out the findings from the Rolla house.

Another specific advantage resulting from the high thermal mass characteristics of underground houses is their ability to maintain a steady or slowly dropping temperature in times of power outages or shortages, thereby preventing damage to plumbing in freezing weather as well as reinforcing a sense of security and independence. Architect John Barnard reports that an owner of an earth sheltered house he designed in Casper, Wyoming, went away for a week in February and turned off the furnace. Although outside temperatures fell well below zero for several stretches of time, the house temperature never dropped below 50°F (10°C).

The thermal mass of earth sheltered structures also permits good integration with other energy systems such as passive solar collection and wood-burning fireplaces that provide heat on a fluctuating basis.

Passive and Active Solar Features

It is no accident that nearly every earth sheltered house featured in this book is designed with an eye toward taking advantage of passive solar gain. The large thermal mass of earth, in combination with the large quantities of masonry generally used to construct such houses, make incorporation of passive solar features a logical—and energy-saving—choice for builders of earth sheltered homes.

Passive solar technology can quite easily be incorporated into standard types of earth sheltered houses, particularly those that have a primarily southern orientation. When converted to heat, the radiant energy from the sunlight admitted to the house through the long banks of south-oriented windows helps heat the entire building. The walls and floors of the structure, which are usually constructed of large quantities of reinforced concrete, act as a large thermal mass. Heat stored in this mass during the day is slowly released at night, thus lessening the need for furnace heating.

Many designers build on the basic concept of passive gain to further enhance thermal performance by using such features as thermal shutters, special drapes, or dark-colored barrels filled with water to hold heat longer.

It is important to note that, while aboveground houses can also be equipped to benefit from passive solar, the large structural mass required for heat storage must be deliberately designed and specially built into these structures. In earth sheltered homes, the large mass is, of necessity, part of the structural design of the house. Few additional costs (i.e., for items such as sunshades) need be incurred in maximizing passive solar gain in an earth sheltered home.

A number of earth sheltered home designers have chosen to boost energy savings even further by using active solar systems. Because earth sheltered structures generally have lower heating requirements than do conventional homes, they require a proportionately lower investment in solar heating collection equipment.

The storage space needed for an active solar system is also reduced in proportion to the heating requirement, and the large thermal

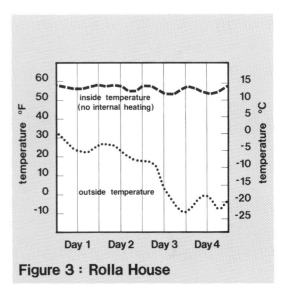

Figure 3 : Rolla House

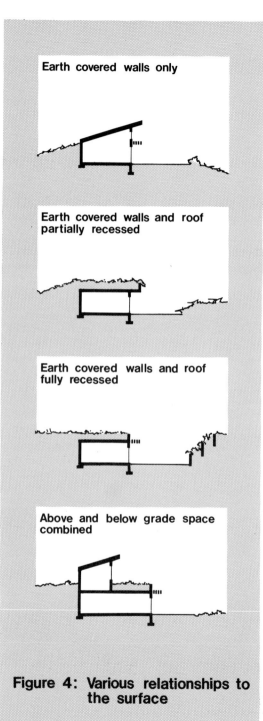

Earth covered walls only

Earth covered walls and roof partially recessed

Earth covered walls and roof fully recessed

Above and below grade space combined

Figure 4: Various relationships to the surface

mass reduces the storage space needs even further. Hence, less floor area is required for the storage medium (e.g., the rock storage bins used in the Seward town houses and Whitewater State Park house and the sand storage bed in the Feuille house). Because of the relatively high costs of active solar systems and the inherently low energy use of earth sheltered structures, however, a payoff in terms of energy savings derived from incorporating active solar features is usually difficult to achieve—even when only a small solar collector (for space heating purposes) is used.

Definitions of *Earth Sheltered House*

Because the use of earth in housing design is a rather general concept, no universally accepted definition of *earth sheltered house* yet exists. Some legal definitions have specified that anywhere from 50 to 80 percent of the roof area must be covered with earth. But a broader definition requiring that 50 percent of the exterior envelope of the building be earth covered allows more latitude in design. For example, the relationship of the house to the ground surface can vary considerably, as shown in the adjacent diagram (Figure 4). In fact, an earth sheltered house may not be below the natural ground level at all, and the roof may or may not be earth covered.

The amount of earth covering and type of structure used to hold back the earth are also subject to considerable variation. Earth sheltering may involve berming only against the walls, while using a conventional well-insulated roof; in a fully earth sheltered design, on the other hand, only the window and door surfaces are not covered with earth. Earth coverings range from an 8- to 10-inch-thick (20 to 25 cm) sod roof to 9 feet (2.7 m) of earth on the roof.

These radically different earth cover alternatives obviously require very different structural systems. To date, the widely varied structural systems used in earth sheltered houses have included poured concrete, concrete

block, precast concrete, post-tensioned concrete, pressure-treated wood, steel highway culverts, and thin shell concrete designs. Regardless of the structural system chosen, structural concerns will always be greater for an earth sheltered house than for a conventional, above-grade house. Because the walls and roofs of earth shelters must be able to support extremely heavy loads, structural calculations for such homes should always be made by a certified structural engineer who is familiar with earth sheltered design.

Basic Designs

Although earth sheltered houses are not limited to any fixed design solutions, the two basic house design concepts illustrated by homes in this book are the elevational and atrium plans.

Elevational House Plans

Elevational designs, which are particularly appropriate for colder climates, group all windows and openings on one exposed elevation (preferably facing south), leaving the three remaining sides earth covered. The already low energy requirements of an elevational structure can often be reduced even further by using south-facing windows to maximize the benefits of passive solar heating.

In a one-story house, the major living and sleeping spaces are nearly always placed along the exposed elevation; secondary spaces not requiring windows (e.g., baths, utility and storage rooms) are located behind them, against the earth-covered walls. Living with one window wall is not an experience unique to inhabitants of earth-sheltered houses and, with the probable increase of passive solar heating in conventional housing, will likely become even more common.

For those who have never visited an elevational-type earth sheltered house, interior conditions and lighting can best be compared with those in a modern apartment or condominium where the living area is backed by the access corridor for all the units. Hence, all the windows are on one wall of the enclosed space (unless the unit occupies a corner position). In most cases, window area in the earth sheltered house will probably exceed that of a typical apartment, since larger windows often will be used to admit maximum solar gain in winter (shading or overhangs are used to keep out the summer sunlight). Skylights or light monitors may also be added toward the rear of elevational plans to admit more light and allow natural ventilation.

The major disadvantage of a one-story elevational plan is that the internal circulation can become rather lengthy—especially for large houses—since the main living spaces are essentially lined up like rooms in a motel. One of several ways designers may alleviate this tendency is through use of a more compact, two-level design.

Atrium House Designs

A courtyard or atrium design is particularly appropriate for a flat site. In this type of design, the habitable rooms cluster around a central courtyard, which provides abundant access to natural light. In its simplest form, the atrium is a square court with living spaces on four sides, although some plans place the living spaces on just three sides, leaving one side open for light, view, and access. Other larger plans may use two or more courtyards. In warmer climates, the atrium area may be used to provide air circulation between rooms; in colder areas, it may be covered with glass.

Advantages associated with this type of plan include the sense of privacy provided by grouping the living space around an interior court and the flexibility with regard to site orientation it provides since—in contrast to most elevational plans—a southerly exposure is not as strongly preferred as a design component.

House Plan Variations

Variations on these two basic house designs include homes that have windows in more than one wall (such as the Earthtech 5 and 6 homes) or combinations of the atrium and elevational types. Houses may have one or two levels; in some cases they may be partially above and partially below grade. Naturally, these plan variations also use different amounts of earth cover, which in turn influences energy performance. Most earth sheltered designs, however, use less energy than conventional houses while creating interior spaces that feel quite similar to completely above-grade houses.

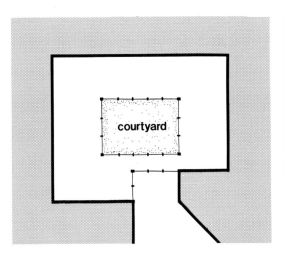

Atrium design – schematic plan

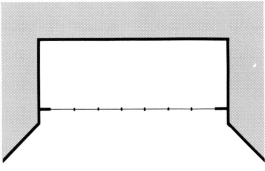

Elevational design – schematic plan

Clark house, Portland, Oregon.

Geier house, Cincinnati, Ohio.

13

Geier house, Cincinnati, Ohio.

Winston house (west side).

Hadley house, Minneapolis, Minnesota.

Winston house, Lyme, New Hampshire.

Landscaping

Whatever type of house design is chosen, landscaping should be a major consideration from the earliest planning stages. It is through appropriate landscaping that one of the primary concepts of earth sheltering—i.e., integration of the structure with its natural surroundings—is accomplished. Several of the houses in this book (e.g., the Remington and Burnsville houses) serve as especially good illustrations of how careful attention to landscaping can aid in integrating the house with the site. Because landscaping is a critical component of the overall design rather than a separate, decorative feature, it must be planned in coordination with all other elements of the house —particularly the structural and waterproofing systems.

Although earth shelter landscaping approaches to date have most often been limited to the use of sods, a number of other approaches—many of which are less costly and require less maintenance than sods—are possible. These include approaches using commercial or special seed mixes, special seed and wildflower mixtures, perennial flowers, drought-tolerant ground covers or vines, commercial shrubs, and native plants. In arid regions, rock gardens may be used to great advantage in successfully blending the site with the building.

Now that many of the initial questions relating to earth sheltered building systems have been answered satisfactorily, designers are increasingly turning to the as yet untapped potential of different landscaping approaches in order to enhance the attractiveness of their earth sheltered homes. Not only do landscaping techniques complete the architectural design; they also assist in the success of waterproofing and insulation systems of underground structures.

Historical Background

Cave Dwellers to 1960

Although most people consider earth sheltering a novel idea, living underground is hardly a twentieth-century phenomenon. From prehistoric times to the present, people all over the world have built and lived below the earth's surface.

Prehistoric cave dwellers, seeking warmth and protection from wild creatures and the elements, chose an existing natural earth form—the cave—that provided those qualities. In fact, the current existence of inhabited cave dwellings in the Loire and Cher Valleys of France provides evidence that, given the proper geology and hydrology, caves can be converted into very comfortable—and extremely private—living spaces.

Throughout history, human beings have often turned to the earth for protection against climatic extremes and danger. Around A.D. 800, the people of Cappadocia, Turkey, carved out underground chambers in spines of soft rock—partially in response to the scarcity of good timber and materials for mortar but mainly to protect the inhabitants from invaders.

For centuries, residents of Matmata, Tunisia, have carved into the soft rock to create atrium houses in which several excavated rooms with 15-foot-high (4.5 m) vaulted ceilings open out onto a single sunken courtyard. These houses are built below ground to protect the inhabitants from the extreme daytime heat and nighttime cold typical of this desert region.

In China, the courtyard-type houses that dot the landscape were dug into the loess soil to combat the hot summers and bitterly cold winters. Farming is carried out on the earth-covered roofs of these houses. In the American Midwest, sod houses and dugouts in the 1800s were also built in response to severe heat and cold, as well as to a lack of building materials and fuel to burn. Sod houses are still in use today in Scandinavian countries.

Given the successful application of underground building technology over the centuries and the effective temperature control and protection that resulted from its use, why did the concept not become more universally applied?

Atrium houses in Matmata, Tunisia.

Underground houses in Cappadocia, Turkey.

Courtyard houses in China.

Midwestern sod house.

The answer is related to the lack of modern construction methods and materials available when most of the aforementioned earth sheltered dwellings were built. Hence, the benefits of these houses were accompanied by drawbacks associated with the ground: dampness, insects and vermin, difficulty keeping them clean, lack of view, and so forth. When conditions changed so that other building materials were readily available and fuel was relatively cheap and easy to obtain, people left their in-ground dwellings for the convenience and status of above-grade homes.

Basements and Basement Houses

In the years since midwesterners abandoned their sod homes, there has been little need in the United States to consider below-ground housing as an option, except as it applies to basements.

Basements, which have been routinely constructed as additions to houses since the early 1900s, are particularly desirable in houses in northern climates, where frost footings have to extend well below ground level. They offer the additional advantage of providing cheap additional space that costs relatively little more to heat or cool than a conventional house without a basement.

Yet, although most people readily acknowledge the usefulness of basements, they also associate basements with undesirable characteristics—usually based on their own living experiences. Because basements traditionally provided cheap, additional space, minimum construction practices were exercised in building them: no reinforcing was applied to limit cracking, no waterproofing (at best, some dampproofing) was installed, and only minimal provisions were made for light and ventilation.

In the past twenty-five years, however, people have increasingly tried to incorporate basements into their homes as real "living space" by finishing off the basement as a family rec room, study, or workshop. Those who have done so are probably aware that adequate lighting, ventilation, waterproofing, and insulation can make the difference between a "musty old basement" and an attractive, comfortable place to work or relax.

Typical waterproofing detail for earth-covered roof (butyl rubber membrane).

To some extent, basements have been associated with negative perceptions people tend to have concerning below-grade housing. After World War II, many families built so-called "basement houses" to live in while they worked on or saved for the remainder of the cost of constructing a house. Often, these families made these "basement houses" their homes for years. Some zoning ordinances enacted in the 1950s to prevent the construction of such unsightly eyesores have hindered construction of earth sheltered homes twenty years later.

Earth Sheltering in the Sixties

Probably the most unusual examples of earth shelters to emerge in the 1960s were a few houses that were built as rather elaborate fallout shelters in response to prevalent fears of an atomic war. In 1962, a full-size example of such a home was built for the World's Fair in Seattle, Washington, where it was toured by thousands of people.

By the late sixties, the fears of atomic war had given way to an increased awareness of the fragility of our environment and ecological systems. Environmentalists touted the concept of earth sheltering, in combination with generous, thoughtful landscaping, as a means of softening the visual and environmental impacts of buildings. Architect Malcolm Wells was a pioneer in this drive toward building without destroying the earth. In 1965 Philip Johnson designed one of the first houses to reflect this environmental concern—an earth-covered house on the edge of a small lake near Cincinnati—primarily for the aesthetic effect of blending it into the surrounding land forms.

Earth Sheltering—1970 to the Present

Environmental and ecological concerns were still the primary reasons for designing with the earth when John Barnard planned the Ecology House in Massachusetts and Don Metz completed his Winston House in New Hampshire in 1972. With the 1973 oil embargo, however, the energy advantages of building underground quickly came to the fore.

Insufficient public awareness and understanding of earth sheltering concepts, coupled with a lack of construction expertise related to this type of building, kept the numbers of such houses quite low, however: by 1976, fewer than fifty truly earth sheltered houses had been built in the entire United States. Over the past several years, both of these obstacles have been overcome to a great extent—the former, through articles and books about earth sheltering, and the latter as contractors and builders have, through direct and sometimes painful experience, developed safe and increasingly cost-effective construction techniques.

Through the sixties and early seventies, a few farsighted, innovative architects continued to design and build earth sheltered houses with little fanfare and without arousing a great deal of interest by either the media or the general public. As energy costs continued their steady climb, however, these houses became the object of increasing attention and curiosity.

The extent of the public's desire for more specific information about earth sheltering became evident early in 1978, when the Underground Space Center published the first basic guide to the concepts and technology of earth sheltering, *Earth Sheltered Housing Design: Guidelines, Examples, and References.*

With some trepidation, the authors ordered an initial printing of 4,000 copies. Much to their surprise, requests for the book began arriving before they had finished writing it, and the first printing sold out in three months. By 1981, more than 160,000 copies of the book had been sold, and sales are still going strong.

Earth Sheltered Housing Design was written as part of a research study commissioned by the Minnesota legislature. Since that project was completed, the Underground Space Center and other research centers have studied earth sheltered solutions to many existing environmental and population problems. Several of these organizations offer academic and/or short courses, conferences, and seminars on various aspects of earth sheltering technology.

Now that a number of the initial uncertainties about structural design, proper waterproofing techniques, and optimal insulation installation have been dealt with through experience (although questions of "ideal" structure and insulation, for example, are by no means fully answered), many researchers are turning their attention to gathering and analyzing reliable data on the thermal performance of earth sheltered structures. The Underground Space Center, Oklahoma State University, and the University of Missouri at Rolla have all been involved in research projects studying energy use by underground homes.

At Texas Tech University, professors have studied the role earth sheltering plays in mitigating the effects of natural disasters and issues related to interior design and consumer acceptance of earth sheltered homes. Both the Underground Space Center and the University of Texas at Arlington have evaluated the impact of earth sheltering technology on community design; the Underground Space Center's book, *Earth Sheltered Community Design*, published in 1981, examines such community developments in depth.

Legislation and Earth Sheltered Homes

On the federal government level, the most important legislation to date concerning earth sheltered housing is the Solar Energy and Energy Conservation Bank Bill, passed by the United States Congress in the summer of 1980. This bill provides low-interest loans for earth sheltered homes, as well as for houses that incorporate passive solar and other energy-conserving features.

Energy agencies in a number of states—including Michigan, California, Wisconsin, Missouri, and Montana—have shown interest in earth sheltered housing. In Minnesota earth sheltering is among the energy features for which home owners can claim state income tax credits; Indiana provides similar tax credits for earth sheltered homes incorporating passive solar design. In addition, seven of the houses described in this book were built with funding provided by the Minnesota legislature for the Earth Sheltered Housing Demonstration Project.

But despite legislation and increasing general awareness of the value of earth sheltering, experts in this field agree that—like many innovative technologies—it is not yet fully accepted by either professionals or the general public. This lack of acceptance is not entirely due to the unique structural requirements of earth sheltered buildings or to negative

psychological reactions from the public. A 1980 study by the Underground Space Center for the Department of Housing and Urban Development (HUD) found that major obstacles to increased construction of such houses have to do with existing financial practices, zoning ordinances, and building code requirements.

As more earth sheltered homes have been built and more information about them has become available, much of the initial skepticism with which earth sheltering concepts were greeted by bankers, other lending institution personnel, code officials, and appraisers—as well as quite a few members of the general public—has gradually given way to an increasing acceptance. But not until enough successful examples of earth sheltered dwellings are built and enough reliable energy data are gathered and analyzed will the severest critics be convinced that earth sheltered houses are not just a passing fad.

The twenty-three homes in this book—homes built for durability, comfort, and energy efficiency—should help dispel that notion. Although these home owners will experience the immediate, direct gains from earth sheltering in the money they will save on fuel bills—as well as the reduced maintenance, quiet, and protection provided by their homes—it is the community at large that will benefit over the long term from the aesthetic and environmental benefits this unique type of architecture provides.

Metz house entrance, Lyme, New Hampshire.

Minnesota Housing Finance Agency Demonstration Houses Part 2

Introduction

The seven MHFA houses provide an excellent introduction to the variety of design options available to those interested in building an earth sheltered residence. The plans range from the basic one-story design of the Camden State Park house to the rather unusual "enclosed atrium" design employed in the Waseca house.

The houses also illustrate ways in which design details—such as wood decking on the Whitewater and Wild River State Park houses, the concrete pipes used as retaining walls in the Camden State Park house, and the utilities shaft of the Burnsville house—can be used to impart a sense of uniqueness to the elevational plan typical of many earth sheltered homes.

In addition, the variety of the site settings—which include urban (Seward town houses), suburban (Burnsville house), small town (Waseca and Willmar), and rural (the three state park houses)—demonstrates that earth sheltering technology can be applied in widely different locales.

In comparing these houses, it is important to note that, despite major differences in aspects of design, detailing, and location, all the homes incorporate passive solar features. The designers' emphasis on taking advantage of the dual benefits to be gained from combining earth sheltering with passive solar technologies is a concept shared with many architects and builders of earth sheltered structures. The importance of using *both* technologies in conjunction in order to achieve maximum aesthetic and energy-conservation goals cannot be overemphasized.

Construction costs (excluding construction financing, land costs, and legal and administrative fees) are summarized for each earth sheltered home built under the MHFA demonstration project. It should be noted that, in general, these costs are somewhat higher (in constant dollars) than would be the case for comparable earth sheltered homes being built today, due in part to special contractual conditions of the project. In addition, these houses were built during a period when the state was experiencing a concrete shortage and a construction boom, contributing to higher costs for both materials and labor. Finally, some of the labor costs undoubtedly reflect the fact that few builders had had experience in using techniques associated with earth sheltered construction at the time these homes were built.

Camden State Park House

The Camden State Park house demonstrates how unique and interesting details can be incorporated into a basic, straightforward design, yielding a house that looks distinctive but is architecturally unpretentious and uncomplicated. Its simplicity also helps the house blend in well with the natural environment of the park. Like the Whitewater State Park house (pages 46-49), the Camden house can accommodate several family sizes and a variety of lifestyles.

Oriented south to take advantage of passive solar gain, the house lies near the park entrance on the edge of a reclaimed 120-acre (48-ha) gravel pit operation. Natural regrowth of the area, starting with cottonwood trees, will be allowed to occur.

Architect Peter Pfister of the Architectural Alliance has replaced traditional retaining walls with large precast concrete pipes at the south corners of the building. In addition to retaining earth for grade changes, the pipes support the trellis above the south-facing windows and give a sculptured look to the face of the house. On the east side of the structure, a larger precast pipe covered with earth serves as the entranceway. By using the concrete pipe, Pfister has designed an attractive entrance without having to completely break the earth berm on that side of the house.

The south-facing bedrooms and living rooms receive low-angle winter sunshine and provide emergency exits. On the northern, interior side of the building are the kitchen, eating, bath, and utility areas. At the back of the house, a wood-frame tower projecting through the roof contains two large skylights that provide light to the rear of the house, solar collectors used for operating an active domestic hot water heating system, and a turbine ventilator for cross-ventilation in summer.

It is expected that the passive solar gain, combined with energy conservation aspects such as earth sheltering and automatic, motor-operated nighttime window insulation on the major glass areas, will significantly reduce the heating demand of the house.

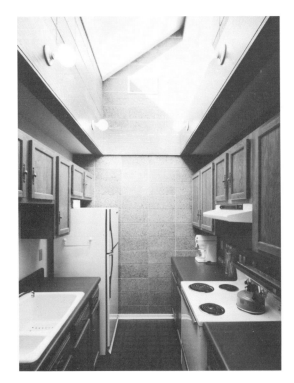

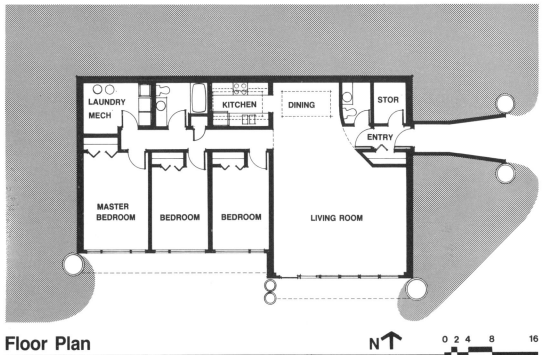

Floor Plan

N↑   0 2 4   8   16

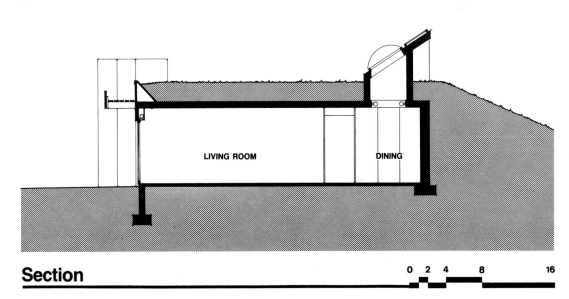

Section

0 2 4   8   16

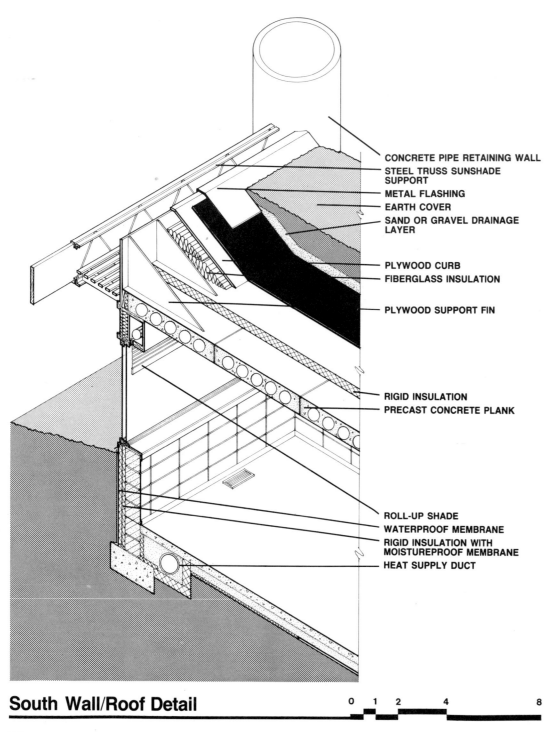

CONCRETE PIPE RETAINING WALL

STEEL TRUSS SUNSHADE SUPPORT

METAL FLASHING

EARTH COVER

SAND OR GRAVEL DRAINAGE LAYER

PLYWOOD CURB

FIBERGLASS INSULATION

PLYWOOD SUPPORT FIN

RIGID INSULATION

PRECAST CONCRETE PLANK

ROLL-UP SHADE

WATERPROOF MEMBRANE

RIGID INSULATION WITH MOISTUREPROOF MEMBRANE

HEAT SUPPLY DUCT

South Wall/Roof Detail

0   1   2   4   8

LOCATION:	Camden State Park, Lynd, Minnesota
ARCHITECT:	Architectural Alliance
ENGINEER:	Bressler Armitage Lunde
CONTRACTOR:	Bladholm and Hess
CONSTRUCTED:	June 1980
GROSS AREA:	1,640 sq. ft. (148 ca)
STRUCTURE:	Reinforced concrete block walls, precast concrete roof, concrete slab-on-grade floor
EARTH COVER:	80% on roof at 18 in. (46 cm) 80% on walls
INSULATION:	Roof—4 in. (10 cm) rigid insulation Walls—4 in. (10 cm) rigid insulation
WATERPROOFING:	Butyl membrane
HEATING DEGREE DAYS:	8,000
HEATING SYSTEM:	Passive solar, wood, electric backup
COOLING SYSTEM:	Natural ventilation

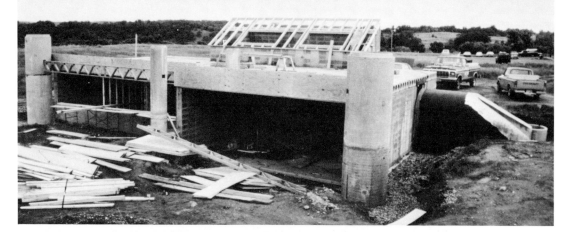

CONSTRUCTION COSTS*:

Site work	$ 4,800
Concrete/masonry	27,340
Waterproofing	15,300
Carpentry	23,780
Heating/plumbing	18,080
Electrical	4,630
Interior finishes	6,700
General requirements	5,545
TOTAL	**$106,175**

*These costs reflect special MHFA Demonstration Project requirements and local construction industry conditions.

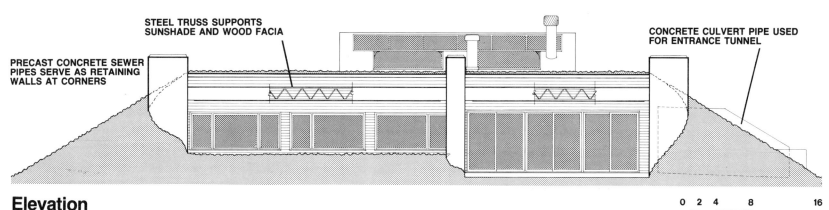

STEEL TRUSS SUPPORTS SUNSHADE AND WOOD FACIA

PRECAST CONCRETE SEWER PIPES SERVE AS RETAINING WALLS AT CORNERS

CONCRETE CULVERT PIPE USED FOR ENTRANCE TUNNEL

Elevation

0 2 4 8 16

Willmar House

Realtors in southwest Minnesota, where this house is located, say that the most frequently asked questions by potential home buyers concern energy costs. Genesis Architecture incorporated earth sheltering and passive solar features into this house with energy conservation in mind.

Located on a south-sloping site, the Willmar house is a good illustration of a typical two-story elevational earth sheltered home design. On the upper level, the main living spaces are grouped around south-facing windows; the bedrooms are located on the cooler lower level. A deck on the east end of the house provides additional view and cross-ventilation.

The designers have maximized the earth cover by using berms up to the window sills on the lower level. A south overhang provides shading for the glass in summer and allows winter sun to penetrate the entire depth of the house in winter.

A forced-air, electric furnace is the primary heating source for the house. Direct passive solar gain, stored in the quarry-tile-on-concrete floor, supplements the furnace heat, along with a fireplace with a heating coil connected to the air return. The south-facing, double-glazed windows have insulated panels on the inside to keep warmth in on cloudy days and at night.

A clerestory provides ventilation and natural light to the rear spaces of the house. At the top of the clerestory skylight space, a fan supplements the natural rate of air flow in the house; set on a thermostat in the kitchen, it operates automatically. Two shafts, containing plumbing vents and exhaust fan vents, penetrate the roof through the clerestory, thus avoiding extra protrusions through the roof of the house.

In addition to the energy savings the house offers, the owner cites quietness and security from severe weather as advantages of this earth sheltered home.

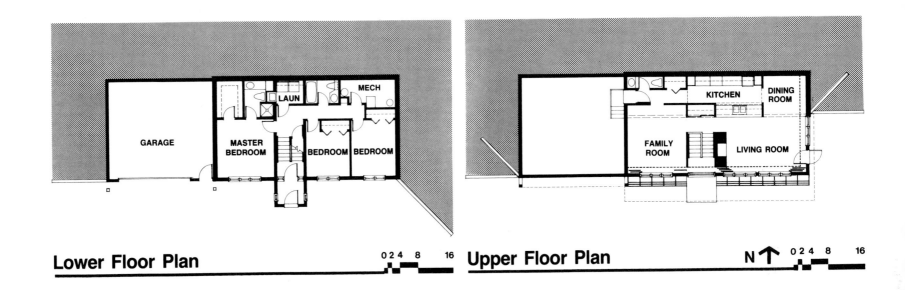

Lower Floor Plan

GARAGE

MASTER BEDROOM

BEDROOM

BEDROOM

LAUN

MECH

0 2 4 8 16

Upper Floor Plan

KITCHEN

DINING ROOM

FAMILY ROOM

LIVING ROOM

N ↑ 0 2 4 8 16

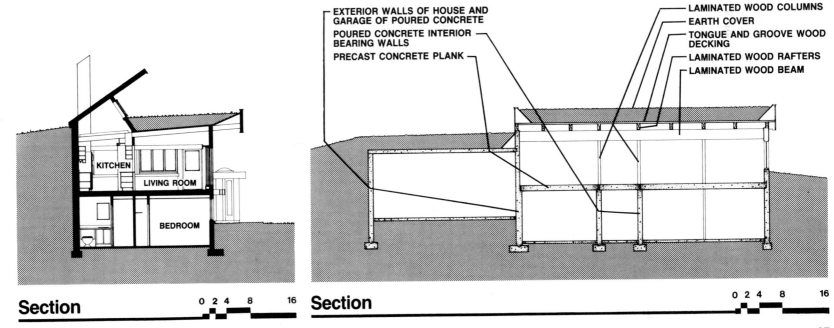

Section

KITCHEN

LIVING ROOM

BEDROOM

0 2 4 8 16

Section

EXTERIOR WALLS OF HOUSE AND GARAGE OF POURED CONCRETE

POURED CONCRETE INTERIOR BEARING WALLS

PRECAST CONCRETE PLANK

LAMINATED WOOD COLUMNS

EARTH COVER

TONGUE AND GROOVE WOOD DECKING

LAMINATED WOOD RAFTERS

LAMINATED WOOD BEAM

0 2 4 8 16

27

LOCATION:	Willmar, Minnesota
ARCHITECT:	Genesis Architecture
ENGINEER:	LWSM
CONTRACTOR:	Willmar Area Vocational Technical Institute, carpentry program
CONSTRUCTED:	1978

GROSS AREA:	2,204 sq. ft. (198 ca)
STRUCTURE:	Reinforced concrete and wood stud walls, laminated wood beams with wood roof decking, precast and concrete slab-on-grade floors
EARTH COVER:	84% on roof at 18 in. (46 cm) 69% on walls
INSULATION:	Roof—4 in. (10 cm) rigid insulation Walls—4 in. (10 cm) rigid insulation on block walls, 6 in. (15 cm) fiberglass batt insulation in stud walls
WATERPROOFING:	Butyl rubber on wood roof, bentonite panels on concrete roof and walls

HEATING DEGREE DAYS:	8,382
HEATING SYSTEM:	Electric furnace, passive solar, wood
COOLING SYSTEM:	Natural ventilation

CONSTRUCTION COSTS*:

Site work	$ 9,390
Concrete/masonry	13,275
Waterproofing	7,170
Metals	2,625
Carpentry	21,430
Heating	1,945
Plumbing	4,040
Electrical	2,865
Interior finishes	7,500
TOTAL	**$70,240**

*These costs reflect special MHFA Demonstration Project requirements and local construction industry conditions.

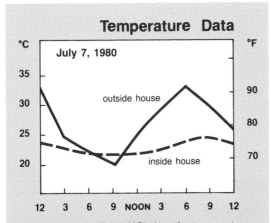

Temperature Data

°C — July 7, 1980 — °F

outside house

inside house

35

30

25

20

90

80

70

12 3 6 9 NOON 3 6 9 12

On a sunny, 92°F (33.3°C) day, the temperature inside this un-air-conditioned house fluctuated 5.4°F (3°C). Temperatures taken from sensors at the upper and lower levels of the house during the same day differed by a maximum of 8.5°F (4.7°C).

Waseca House

The Design Consortium set out to design an energy-efficient home that would not require the south-sloping orientation and linear room arrangement typical of many earth sheltered houses. The result is an "enclosed-atrium"-type home that is distinctly different from the majority of designs associated with earth sheltered architecture.

Like most atrium designs, the Waseca house is internally oriented, with rooms organized around a central light source, rather than overlooking an exterior landscape. The 1,300-square-foot (90-ca) living area is, however, considerably smaller than that of most atrium-type earth sheltered homes. This house

demonstrates that an atrium plan need not necessitate more space than other types of earth sheltered house designs. In this case, the designers chose to use the atrium space in a unique and economical way—as a major living area.

The floor plan is organized so that the living/dining room is the hub of activity for the occupants. At the four corners of this central space are the smaller, more private living areas such as bedrooms and family rooms. These are connected to the front and rear courtyards by sliding glass doors, allowing natural light and ventilation to penetrate the interior of the house. The mechanical, storage, and bathroom

spaces, which do not require exterior views, are located away from the naturally lighted areas of the house.

Operable clerestory windows in the raised roof atrium area permit light to penetrate to the major living space in cooler months and provide ventilation in summer. By opening the courtyards at both ends, this area benefits from dual exterior exposure and thorough cross-ventilation. Massed evergreen plantings sheltering the rear courtyard increase the occupants' sense of privacy while reflecting the natural surroundings.

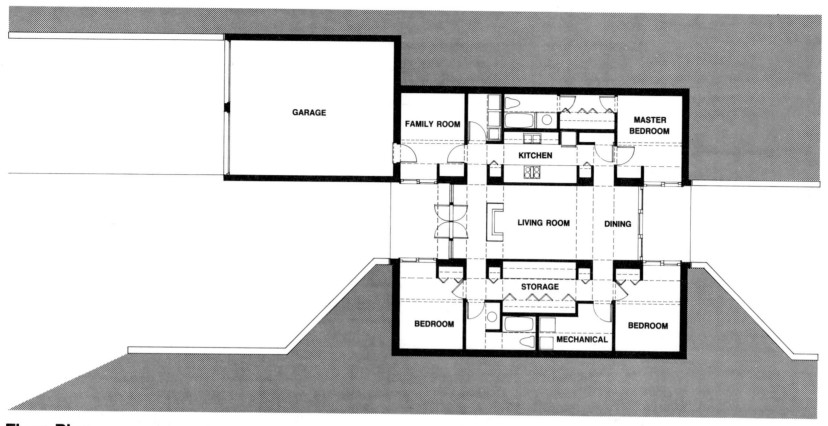

GARAGE
FAMILY ROOM
MASTER
BEDROOM
KITCHEN
LIVING ROOM
DINING
STORAGE
BEDROOM
MECHANICAL
BEDROOM
0 2 4 8 16

Floor Plan

ENTRY
LIVING ROOM
COURT
0 2 4 8 16

Section

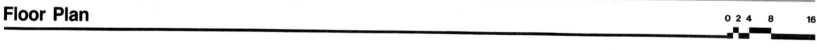

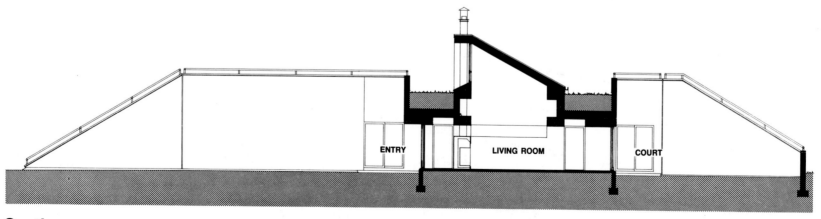

LOCATION:	Waseca, Minnesota
ARCHITECT:	Design Consortium, Inc.
ENGINEER:	Nelson/Rudie Associates
CONTRACTOR:	Associated Lumber Marts
CONSTRUCTED:	1980
PHOTOGRAPHY:	©1980, Phillip MacMillan James
GROSS AREA:	1,300 sq. ft. (117 ca)
STRUCTURE:	Reinforced concrete block walls, precast concrete planks with sloped concrete topping roof, concrete slab-on-grade floor
EARTH COVER:	90% on roof at 24 in. (61 cm) 76% on walls
INSULATION:	Roof—3 in. (8 cm) rigid insulation Walls—2 in. (5 cm) rigid insulation
WATERPROOFING:	Bentonite
HEATING SYSTEM:	Gas forced-air furnace, electric backup, heat-circulating fireplace
COOLING SYSTEM:	Natural ventilation

CONSTRUCTION COSTS*:

Site work	$ 7,635
Concrete/masonry	37,820
Metals	1,230
Waterproofing	5,985
Carpentry	53,060
Heating/plumbing	6,100
Electrical	12,090
Interior work	15,730
TOTAL	**$139,650**

*These costs reflect special MHFA Demonstration Project requirements and local construction industry conditions.

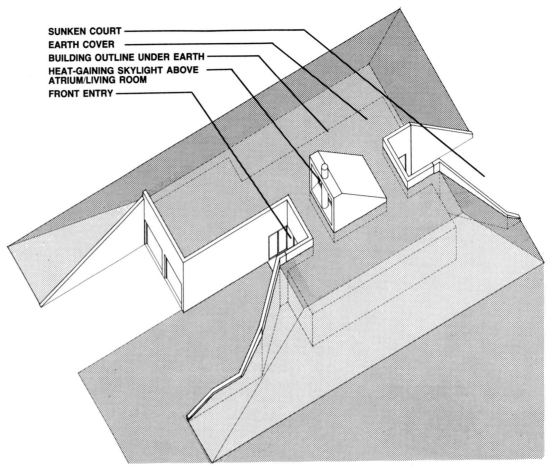

SUNKEN COURT
EARTH COVER
BUILDING OUTLINE UNDER EARTH
HEAT-GAINING SKYLIGHT ABOVE ATRIUM/LIVING ROOM
FRONT ENTRY

Aerial View (axonometric)

0 4 8 16

Burnsville House

This suburban Minneapolis home was designed to meet three objectives: conservation of energy by using earth sheltering and insulation in conjunction with passive solar heating, reasonable construction costs, and integration of energy efficiency and cost effectiveness with an aesthetically pleasing design that uses the earth cover as a positive design feature.

A major factor influencing the design of the Burnsville house was the challenge presented by the steeply sloping, heavily wooded site. Working with the slope, the designers placed the house into an east-west ridge that slopes away to the north and south.

By opening the house to both the north and south sides, the architects have separated the more public entry and driveway area (on the north) from the private view of the outdoors, which faces onto the primary living spaces. The original slopes to the site have been maintained so that the house blends in very naturally with the surrounding land forms. Treated timber retaining walls and planters harmonize with the rough cedar siding used on the exterior, which in turn has weathered to reflect the natural colors of the woods.

An outdoor deck—placed at the east end of the house so as not to interfere with the sunlight to the lower level—provides an exit to grade from the living areas, which are located on the upper level. In addition to reducing the area required for the exterior building envelope, the compact two-story configuration of the house minimizes the potential problem of lengthy internal circulation.

The roof of the house is completely covered with earth, as are most of the east, west, and north walls. The designers made significant use of retaining walls to manipulate the earth for the elevational changes that were required to achieve earth sheltering to the roof of the structure.

All the utilities have been consolidated in a shaft that extends up the middle of the south-facing front of the house. By means of this unique design feature, roof penetrations through the waterproofing and concrete roof plank have been avoided.

The sloped roof and rather narrow plan allow greater penetration of sunlight into the living spaces. Additional solar heat gain, as well as substantial natural light, is admitted through the large clerestory windows. The living spaces on the upper floor act as collectors for direct passive solar gain. Solar radiation is absorbed by the dark brown, unglazed ceramic tiles of the intermediate floor, which then release heat into spaces on both levels of the house.

In summer, sunshades over the windows, in combination with the tall deciduous trees on the south side of the house, keep out direct sun. The north entry and operable clerestory windows help provide natural ventilation throughout the house.

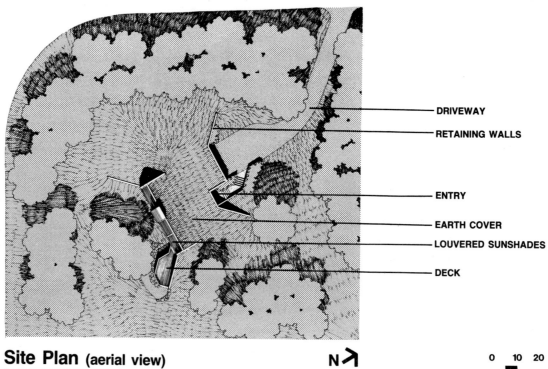

Site Plan (aerial view) N↗ 0 10 20

- DRIVEWAY
- RETAINING WALLS
- ENTRY
- EARTH COVER
- LOUVERED SUNSHADES
- DECK

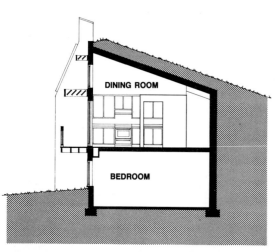

Section 0 2 4 8 16

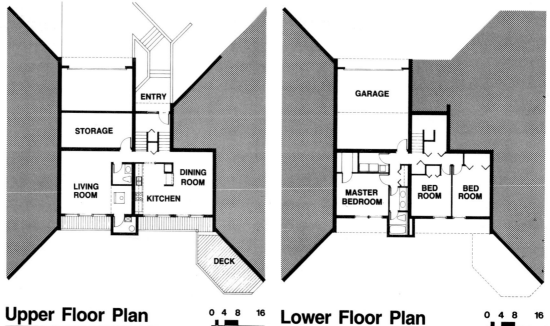

DINING ROOM

BEDROOM

ENTRY

STORAGE

LIVING ROOM

KITCHEN

DINING ROOM

DECK

GARAGE

MASTER BEDROOM

BED ROOM

BED ROOM

Upper Floor Plan 0 4 8 16

Lower Floor Plan 0 4 8 16

LOCATION: Burnsville, Minnesota
ARCHITECT: John Carmody, Tom Ellison
STRUCTURAL ENGINEER: Martin Lunde
MECHANICAL ENGINEER: Terry Tillman
CONTRACTOR: Ellison Design & Construction
CONSTRUCTED: 1979
PHOTOGRAPHY: Tom Ellison

GROSS AREA: 1,950 sq. ft. (176 ca)
STRUCTURE: 12 in. (30 cm) reinforced concrete block and 2 × 6 wood stud walls, precast concrete plank roof, precast and concrete slab-on-grade floors
EARTH COVER: 100% on roof
60% on walls
INSULATION: Roof—6 in. (15 cm) rigid insulation
Walls—4 in. (10 cm) rigid insulation, tapering to 1 in. (2.5 cm)
WATERPROOFING: Butyl rubber membrane

HEATING DEGREE DAYS: 8,382
HEATING SYSTEM: Electric furnace, forced-air system
COOLING SYSTEM: Natural

CONSTRUCTION COSTS*:

Site work	$17,010
Concrete/masonry	17,430
Waterproofing	7,310
Carpentry	40,715
Heating	1,790
Plumbing	2,790
Electrical	2,590
Interior finishes	7,290
General requirements	3,000
TOTAL	**$99,925**

*These costs reflect special MHFA Demonstration Project requirements and local construction industry conditions.

Temperature Data

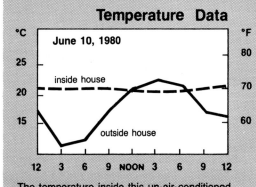

June 10, 1980

inside house

outside house

12 3 6 9 NOON 3 6 9 12

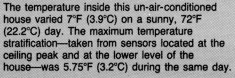

The temperature inside this un-air-conditioned house varied 7°F (3.9°C) on a sunny, 72°F (22.2°C) day. The maximum temperature stratification—taken from sensors located at the ceiling peak and at the lower level of the house—was 5.75°F (3.2°C) during the same day.

Seward Town Houses

One of the most innovative and unusual earth sheltered housing projects to date, the Seward town house development resulted from a cooperative effort involving the community and the architects. The site, located immediately adjacent to a very busy section of freeway and adjoining a major intersection, had become undesirable to most residential developers. It was slated for use by a major restaurant chain when Seward West Redesign, a nonprofit neighborhood corporation concerned about increasing commercialization of the area, proposed an alternative: an earth sheltered residential complex.

The town houses demonstrate how a thoughtful, well-planned design can turn normally undesirable site characteristics to advantage through the application of earth sheltering and passive solar techniques. For example, the fact that the noisy freeway is located immediately north of the site dictated that the complex face south—the ideal orientation for passive solar gain. In addition, by facing the units south and creating a berm of earth over the north end and both sides of the complex, the architects successfully "dampened" the freeway noise.

The twelve-unit (nine two-bedroom, three three-bedroom) development is completely covered by berms on the three sides; the roof is planted with long natural grasses. The north berm, designed as a continuation of the grassy edge predominant along the freeway, is punctuated by entrances to each of the units. On the south are located the primary entrances and the individual unit courtyards where owners may plant gardens or shrubs.

To make the town house units as energy-efficient as possible, the architects incorporated both an active solar system and passive solar features in the design. The active solar system consists of interconnected forced-air flat-plate collectors atop the roof (seven panels for each two-bedroom unit, nine panels for the three-bedroom units) capable of preheating the domestic water and storing excess heat in a rock storage box in the mechanical room.

The south orientation of the windows, combined with manually operated exterior insulating rolling shades, contributes to a significant passive solar gain. In addition, the hollow precast floor acts as a warm air plenum, as fans distribute heat evenly by forcing air through the cores of the floor.

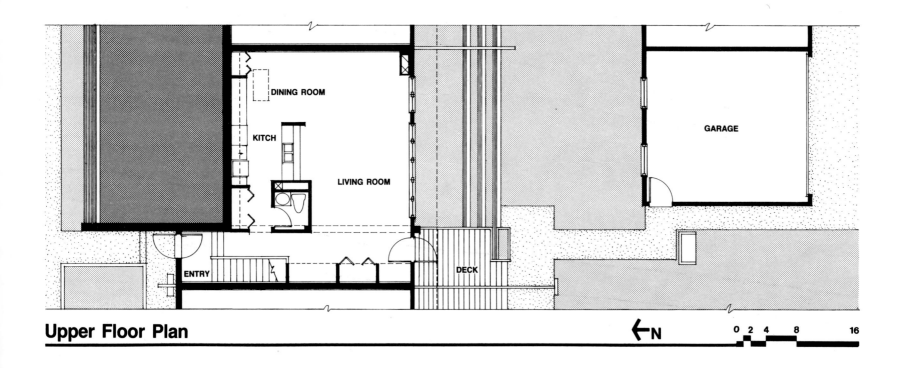

Upper Floor Plan

←N   0 2 4   8   16

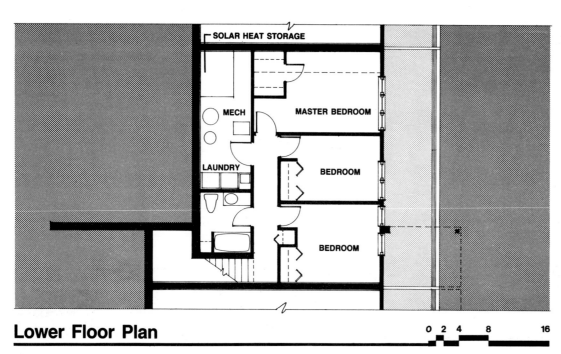

Lower Floor Plan

0 2 4   8   16

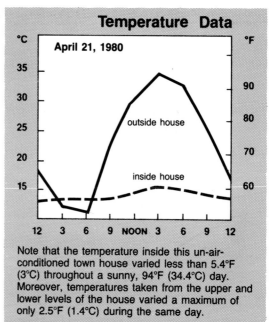

Temperature Data

April 21, 1980

outside house

inside house

Note that the temperature inside this un-air-conditioned town house varied less than 5.4°F (3°C) throughout a sunny, 94°F (34.4°C) day. Moreover, temperatures taken from the upper and lower levels of the house varied a maximum of only 2.5°F (1.4°C) during the same day.

Retaining Wall Detail

- CONTINUOUS TREATED TIMBER DEADMAN WITH EYEBOLTS
- TIE-BACK CABLES
- 2:1 SLOPE ON BERM— SWALE AT BOTTOM
- COMPACTED GRANULAR BACKFILL
- TREATED TIMBER HELD TOGETHER WITH 10-IN. GALVANIZED STEEL SPIKES
- GALVANIZED STEEL EYEBOLTS
- ORIGINAL GRADE
- SIDEWALK

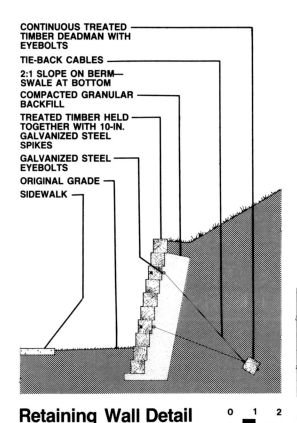

0 1 2

Skylight Detail

- PRECAST CONCRETE ROOF PLANKS
- CONTINUOUS FLASHING FROM SKYLIGHT CURB TO GRADE BELOW
- WATERPROOF MEMBRANE
- WOOD CURB
- POLYESTER BOARD NAILED AND GLUED TO INSULATION AND 2 X 2 FURRING STRIPS
- SKYLIGHT
- RIGID INSULATION

0 1 2

LOCATION:	Minneapolis, Minnesota
ARCHITECT:	Michael Dunn/Close Associates, Inc.
ENGINEER:	Meyer, Borgman & Johnson, Inc.
CONTRACTOR:	Kraus-Anderson of St. Paul
CONSTRUCTED:	1979
PHOTOGRAPHY:	Jerry Mathiason
GROSS AREA:	13,746 total sq. ft. (1,237 ca) 1,387 sq. ft. per three-bedroom unit (3) 1,065 sq. ft. per two-bedroom unit (9)
STRUCTURE:	Reinforced concrete block and wood stud walls, precast concrete roof, precast panel and slab-on-grade floors
EARTH COVER:	100% on roof at 18 in. (46 cm) 95% on walls
INSULATION:	Roof—4 in. (10 cm) rigid insulation Walls—3 in. (8 cm) rigid insulation on concrete block walls, 6 in. (15 cm) fiberglass batt insulation in wood stud walls
WATERPROOFING:	Sprayed-on bentonite
HEATING DEGREE DAYS:	7,906
HEATING SYSTEM:	Gas forced-air furnace, active and passive solar
COOLING SYSTEM:	Natural ventilation
CONSTRUCTION COST*:	$783,815

*These costs reflect special MHFA Demonstration Project requirements and local construction industry conditions.

Section

- SOLAR COLLECTORS

KITCHEN   LIVING ROOM

BEDROOM

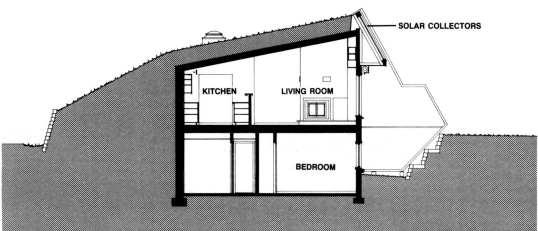

0 2 4 8 16

Wild River State Park House

Both the construction methods and building materials for the Wild River State Park house were deliberately limited—to techniques familiar to and commonly used by builders of small rural houses, and to materials generally available in lumber yards of small communities. The architects imposed these limitations because of the relative remoteness of the house site from a large urban area and in order to demonstrate that an energy-efficient house that uses passive solar collection and storage techniques can be constructed with materials readily available to the general public.

Set into a south-facing hillside with the south wall oriented 15 degrees east of south, the Wild River house benefits from passive solar gain. The entire north wall, most of the west wall, and portions of the east and south

walls are earth sheltered. The existing soil—a sand gravel mix—is ideal for drainage and lessened the problems of waterproofing the house.

On the east end of the upper level, a large deck provides occupants with additional space and a view of the natural park surroundings. A 4-foot-wide (1.2-m) deck along the south side of the house also serves as a summer sun shade for the lower-level windows.

Between the north, sloped ($^5/_{12}$ pitch) half of the roof and the flat, south half of the roof are manually operable clerestory windows. These provide natural light for the rooms at the back of the upper floor of the house and in summer ventilate the high ceiling spaces that normally collect and stratify hot air. The clerestory windows are equipped with manually

operated, insulated shutters that are closed during winter nighttime hours and on cloudy winter days to reduce heat loss. Well integrated with the shapes and materials used in the house, the shutters appear to be almost a part of the structure when they are in the open position and are unobtrusive when closed. Roof overhangs at both the sloped and flat roofs are designed to shade glass areas and to allow all midwinter sunlight to enter the house.

To eliminate stratification of hot air at the top of the high, sloped ceiling on the upper level, a circulating fan draws air through slots in the fins between the clerestory windows. The air is then directed through ducts into a rock bin located below the floor of the lower-level family room and is returned to the forced-air furnace.

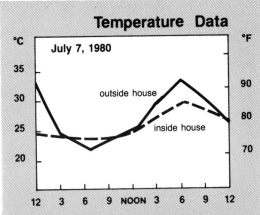

Temperature Data

July 7, 1980

outside house

inside house

Throughout this sunny, 92°F (33.3°C) July day, the interior temperature of this house fluctuated 14°F (7.8°C). The maximum temperature differential between the upper and lower levels of the house during the same day was 21°F (11.7°C).

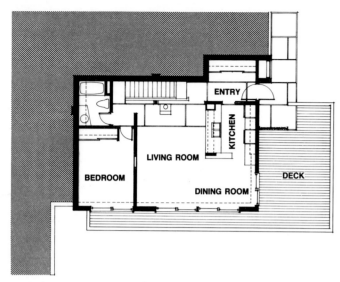

Upper Floor Plan 0 2 4 8 16

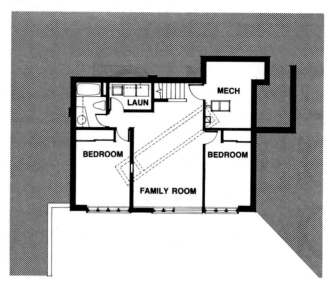

Lower Floor Plan 0 2 4 8 16

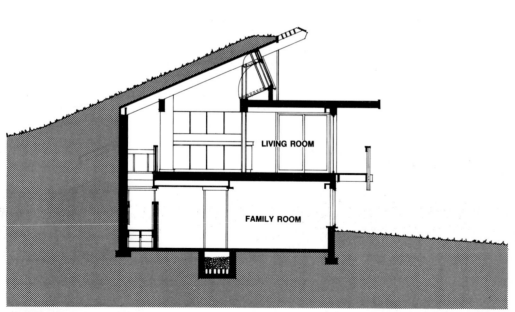

Section 0 2 4 8 16

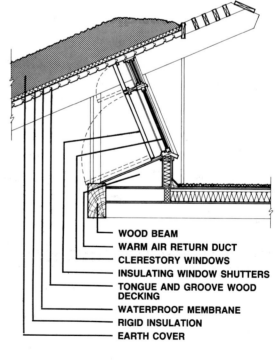

WOOD BEAM
WARM AIR RETURN DUCT
CLERESTORY WINDOWS
INSULATING WINDOW SHUTTERS
TONGUE AND GROOVE WOOD DECKING
WATERPROOF MEMBRANE
RIGID INSULATION
EARTH COVER

Clerestory Detail 0 1 2

LOCATION:	Wild River State Park, Chisago County, Minnesota
ARCHITECT:	McGuire, Engler, Davis Architects
STRUCTURAL ENGINEER:	Bredow Associates, Inc.
MECHANICAL ENGINEER:	Chasney Associates
CONTRACTOR:	Herb Larson Construction
CONSTRUCTED:	Spring 1980
GROSS AREA:	1,928 sq. ft. (174 ca)
STRUCTURE:	Wood stud and reinforced concrete block walls; wood beam and deck roof; wood joist and concrete slab-on-grade floors
EARTH COVER:	68% on roof at 18 in. (46 cm) 50% on walls
INSULATION:	Roof—2 in. (5 cm) rigid insulation on sloped roof, 6 in. (15 cm) fiberglass batt insulation on flat roof Walls—2 in. (5 cm) rigid insulation on exterior of concrete block, 6 in. (15 cm) fiberglass batt insulation in stud wall
WATERPROOFING:	Roof—butyl membrane Walls—troweled-on mastic
HEATING DEGREE DAYS:	8,159
HEATING SYSTEM:	Wood-fired forced-air furnace, electric coils in supply ducts, wood stove
COOLING SYSTEM:	Natural ventilation

CONSTRUCTION COSTS*:

Site work	$ 7, 820
Concrete/masonry	15,220
Waterproofing	6,970
Carpentry	48,420
Heating	5,900
Plumbing	3,700
Electrical	2,500
Interior finishes	9,100
General requirements	9,525
TOTAL	**$109,155**

*These costs reflect special MHFA Demonstration Project requirements and local construction industry conditions.

Whitewater State Park House

In designing this park manager's residence, Close Associates wanted to provide a space arrangement flexible enough to comfortably house various family sizes—an important consideration due to the high turnover among park managers. Additionally, the design had to provide both a public reception area and privacy for the residents. The first requirement was met by including three bedrooms that could be used alternatively as a study or den. The second concern was resolved by designing an entry vestibule roomy enough to accommodate a desk and several visitors but separated from the family living area by glass doors.

Sited at an angle to a gentle, southwest slope, the house faces predominantly south to benefit from solar heat gain. The hill wraps around the structure on three sides, with a break for the garage and main entry.

Above the entry area and stairway, a high vaulted ceiling projects through the earth-covered roof, providing natural light from dome-shaped skylights. The roof of this clerestory is sloped at a 45-degree angle in order to maximize the benefits of rooftop solar collector panels, which are used exclusively for preheating the water supply. Two other skylights supply additional natural light to the kitchen and living areas.

The active solar system consists of two features: the rooftop solar panels, and vertical panels on the south wall which collect heat that is either directly used to heat the house or stored in a rock bin until needed. Unlike the roof collectors, which are used the year-round, the vertical solar collectors operate only in winter.

Passive solar benefits are obtained via the south-facing glazing, which allows the sun to strike and warm the hollow precast concrete floor. By permitting air flow in all directions, the precast panels approximate a warm air plenum through which air is continuously forced by the furnace fan—thus evenly distributing the solar heat. The masonry mass of the house also stores heat from passive gain.

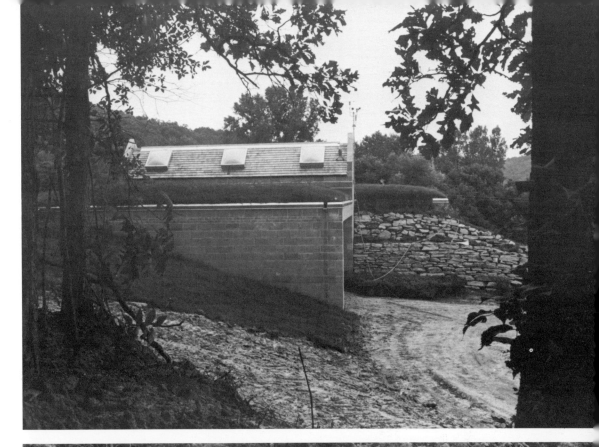

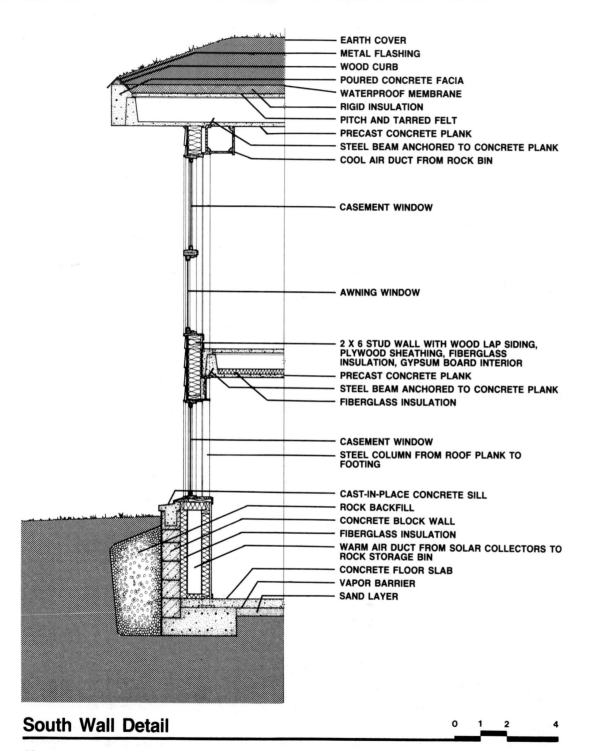

EARTH COVER
METAL FLASHING
WOOD CURB
POURED CONCRETE FACIA
WATERPROOF MEMBRANE
RIGID INSULATION
PITCH AND TARRED FELT
PRECAST CONCRETE PLANK
STEEL BEAM ANCHORED TO CONCRETE PLANK
COOL AIR DUCT FROM ROCK BIN

CASEMENT WINDOW

AWNING WINDOW

2 X 6 STUD WALL WITH WOOD LAP SIDING, PLYWOOD SHEATHING, FIBERGLASS INSULATION, GYPSUM BOARD INTERIOR
PRECAST CONCRETE PLANK
STEEL BEAM ANCHORED TO CONCRETE PLANK
FIBERGLASS INSULATION

CASEMENT WINDOW
STEEL COLUMN FROM ROOF PLANK TO FOOTING

CAST-IN-PLACE CONCRETE SILL
ROCK BACKFILL
CONCRETE BLOCK WALL
FIBERGLASS INSULATION
WARM AIR DUCT FROM SOLAR COLLECTORS TO ROCK STORAGE BIN
CONCRETE FLOOR SLAB
VAPOR BARRIER
SAND LAYER

South Wall Detail

0   1   2   4

LOCATION:	Whitewater State Park, St. Charles, Minnesota
ARCHITECT:	Close Associates, Inc.
STRUCTURAL ENGINEER:	Meyer, Borgman & Johnson, Inc.
MECHANICAL ENGINEER:	John Borry—backup heating system
	Solar Architect/ Engineering—active solar collector systems
CONTRACTOR:	Wolter Lumber Company
CONSTRUCTED:	July 1980
GROSS AREA:	1,891 sq. ft. (170 ca)
STRUCTURE:	Reinforced concrete block walls, Nilcon precast panel roof, Nilcon precast panel and slab-on-grade floor
EARTH COVER:	100% on roof at 18 in. (46 cm) 75% on walls
INSULATION:	Roof—4 in. (10 cm) rigid insulation
	Walls (above grade)—4 in. (10 cm) fill in cavity of west wall; 1½ in. (4 cm) rigid insulation elsewhere
WATERPROOFING:	Built-up coal-tar bitumen and tarred felt

HEATING DEGREE DAYS:	7,850
HEATING SYSTEM:	Electric forced-air furnace, active and passive solar
COOLING SYSTEM:	Natural ventilation

CONSTRUCTION COSTS*:

Site work	$ 5,625
Concrete/masonry	27,560
Metals	2,085
Waterproofing	6,840
Carpentry	23,755
Heating/plumbing	20,405
Electrical	3,985
Interior finishes	5,625
General requirements	3,160
TOTAL	**$99,040**

*These costs reflect special MHFA Demonstration Project requirements and local construction industry conditions.

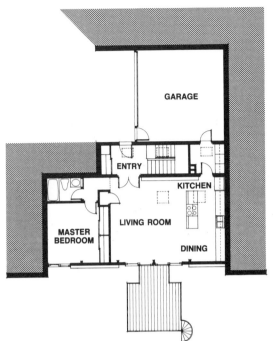

Upper Floor Plan N 0 2 4 8 16

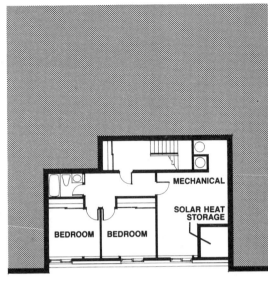

Lower Floor Plan 0 2 4 8 16

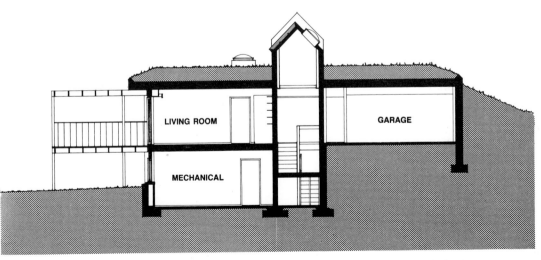

Section 0 2 4 8 16

Sixteen Houses from the U.S. and Europe Part 3

Introduction

In terms of structure, house plan type, size, climate, and location, the sixteen houses in this section illustrate the variety of earth sheltered house designs and configurations possible in a range of climates.

Most of these houses use conventional flat-roofed structural systems that include concrete walls with a precast or poured concrete roof. Others, however, involve more unusual structural systems—Terra-Dome's dome-shaped modules, Architerra's curvilinear rear wall, and the wood roofs of the Earthtech, Feuille, and Suncave houses, for example. The curved retaining walls of the Suncave and Remington houses and the absence of retaining walls in the Hadley house are other variations on the typical structural design of earth sheltered houses.

Reflecting the popularity of the elevational type of house design, many of the houses in this section are elevational structures set into south-facing slopes. A number of other configurations are also included, pointing up the versatility of basic earth sheltered designs. The Clark residence is an atrium-type design, the

Sticks & Stones and Wheeler houses combine atrium with elevational features, and the Earthtech 5 and 6 houses expose the west as well as south walls—a variation on the elevational design.

Like the MHFA demonstration houses, nearly all the homes in this section employ features such as glazing on south walls and massive construction materials in order to maximize passive solar gain potential associated with earth sheltered structures. An exception, the Moreland house in Texas demonstrates that earth sheltering can also prove advantageous in terms of providing passive cooling benefits. Several architects have included active solar features in their houses to reduce energy costs even further. The SunEarth home, for example, combines a number of passive and active solar features to achieve a high degree of energy efficiency.

While the majority of these houses are located in rural or suburban settings, where zoning ordinances and building codes are generally less troublesome than is true with

urban sites, the Clark and Sticks & Stones homes testify to the ability of earth sheltered houses to complement and blend with conventional home styles in typical urban areas.

Moreover, the Architerra development, like the Seward town houses built under the MHFA demonstration project, provides evidence that concepts of earth sheltering can be very successfully used in creating multiresidential complexes, often on sites considered unsuitable for residential construction. The economic and environmental benefits possible through such residential developments cannot—and should not—be overlooked by architects and builders.

The contrasts in style and appearance among these houses, as well as the MHFA demonstration homes, serve to underscore a point frequently emphasized by proponents of earth sheltering—namely, that this type of architecture permits designers considerable flexibility and ingenuity in creating comfortable, attractive, safe, and energy-efficient homes.

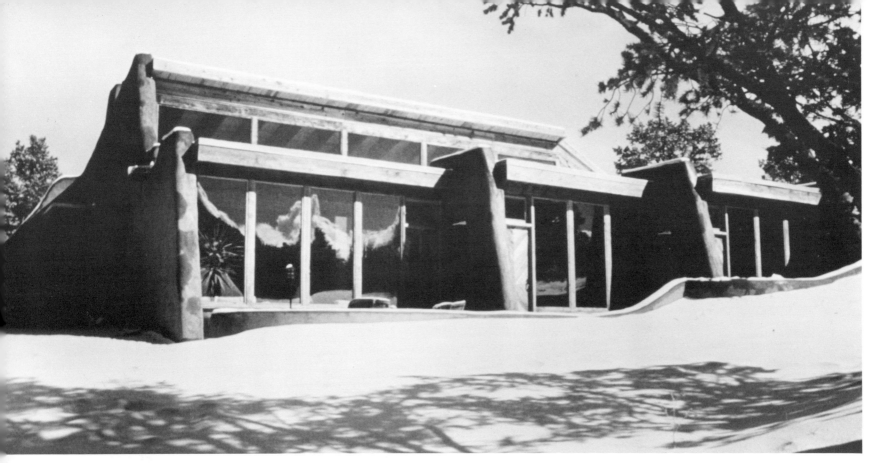

Suncave

Dug into the side of a southwest-facing hill and partly covered by earth on the roof, Suncave was the first passive solar adobe home built on speculation. The earth sheltered design was a response to the sloping site characteristics, a request for low visibility from the neighboring road, and the owners' desire for a high degree of thermal integrity, achievable through earth sheltering. Its earth-covered form permits the house to harmonize with the surrounding terrain, minimize heat loss, and take maximum advantage of solar exposure.

This home is an excellent example of how earth sheltering permits integration of a house with its environment. The earth tone of the brown adobe block used for the west exterior wall (and for all interior mass walls) matches the color of the surrounding soil, thus helping the house blend into the surrounding hill.

Through the use of a curved retaining wall, the architects have created a gently sloping earth berm that echoes naturally occurring forms of the surrounding landscape. In addition to giving the house facade a three-dimensional appearance rather uncommon among earth sheltered houses, the archlike shape of the wall helps effectively resist lateral earth forces. By enclosing the space on the south side of the house, which is used as a patio, the curved wall ensures privacy while adding an outdoor living space for the occupants.

Like the Wild River State Park house (pages 42-45), Suncave features a clerestory element and a partially earth-covered sloping roof. The clerestory windows provide solar gain and natural light to the rooms at the back of the house, as well as assisting with natural ventilation. Two skylights penetrate the north sloping roof over the dining/kitchen area, permitting spot lighting over the dining table and kitchen work space.

The reinforced concrete masonry walls, adobe walls, and floors constitute the main thermal mass of the house—a mass sufficient to keep the house comfortable without backup heating for several days in a climate where outside temperatures reach -15°F (-26°C). The sun provides about 86 percent of the required annual heating; less than a cord of wood has been needed annually to maintain an interior temperature range of 65°-75°F (18°-24°C).

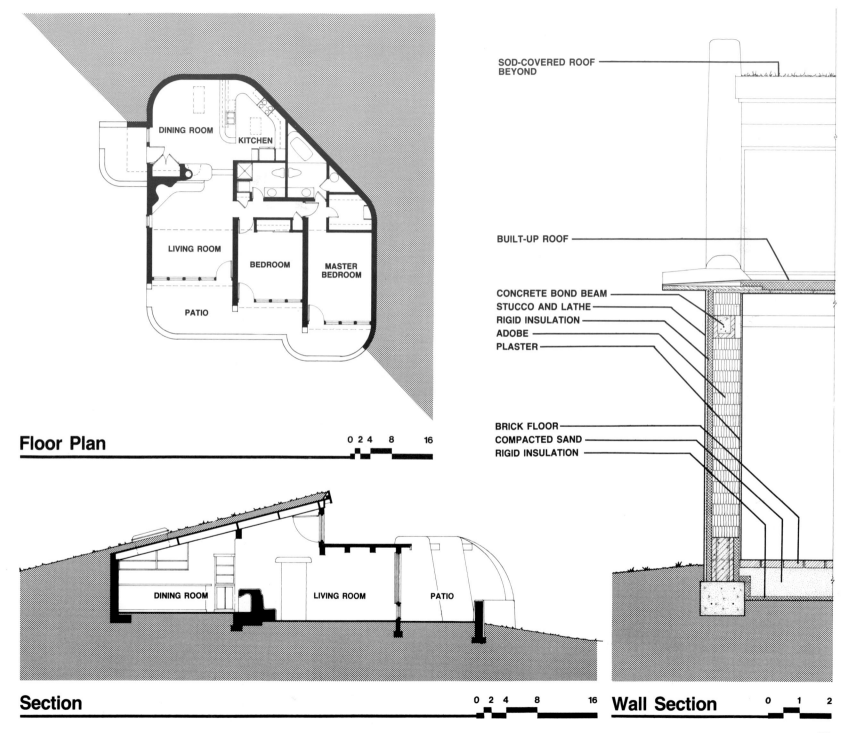

Floor Plan

DINING ROOM

KITCHEN

LIVING ROOM

BEDROOM

MASTER BEDROOM

PATIO

0 2 4 8 16

Section

DINING ROOM

LIVING ROOM

PATIO

0 2 4 8 16

SOD-COVERED ROOF BEYOND

BUILT-UP ROOF

CONCRETE BOND BEAM
STUCCO AND LATHE
RIGID INSULATION
ADOBE
PLASTER

BRICK FLOOR
COMPACTED SAND
RIGID INSULATION

Wall Section

0 1 2

LOCATION: Santa Fe, New Mexico
ARCHITECT: David Wright
ENGINEER: David Wright
CONTRACTOR: Karen Terry
CONSTRUCTED: 1976

GROSS AREA: 1,500 sq. ft. (135 ca)
STRUCTURE: Concrete masonry and adobe walls, wood beam and decking roof, brick on sand floor
EARTH COVER: 60% on roof at 6 in. (15 cm) 50% on walls
INSULATION: Roof—2 in. (5 cm) rigid insulation, 6 in. (15 cm) fiberglass batt insulation Walls—2 in. (5 cm) rigid insulation
WATERPROOFING: Four-ply, 90-lb. built-up roofing

HEATING
DEGREE DAYS: 4,292
HEATING SYSTEM: Passive solar, adobe heat retention fireplaces
COOLING SYSTEM: Natural ventilation

Remington House

In merging gracefully with its natural wooded site, the Remington house takes on a three-dimensional appearance rather unusual among earth sheltered homes. This type of facade, like that of the Suncave dwelling (pages 54-57), is largely due to the curved retaining wall that wraps around the front of the house, creating a private outdoor courtyard.

Another interesting aspect of this house is the relatively large area covered by earth: the ratio of exposed wall area to square footage is quite low. All of the roof and most of the walls are earth covered. The key to this plan is the 48-foot (14.5-m) continuous skylight—called a "solar attic" by architect Richard Webster—that extends the length of the house and provides

substantial natural light throughout the interior spaces. The skylight unit also acts as a solar heat sink from which warm return air is circulated into the rest of the house by the air handler of a split heat pump system. In addition, the unit was designed to accommodate eighteen linear concentrating solar collectors—not yet installed—to run a domestic hot water system.

Even without the active solar system, the house has proven quite energy-efficient, based on temperature monitoring figures and heating costs. The passive solar gain via the skylight is absorbed by the thermal mass of the exposed split-face concrete block, which has been used for both exterior and interior walls.

Inside temperatures were continuously recorded over a four-season cycle the year that the house was completed. Over that one-year period, the highest temperature recorded was 83°F (28°C); the lowest—recorded at Christmas in 1978, when all systems in the house had been deliberately turned off for four days—was 54°F (12°C). The annual inside temperature averaged 71°F (22°C) over the same period, when the average outside temperature was 49°F (9°C).

For the first winter, the total heating costs for the all-electric structure were $180: $75 for two cords of wood, $10 for coal, and $95 for electricity (above basic appliance use).

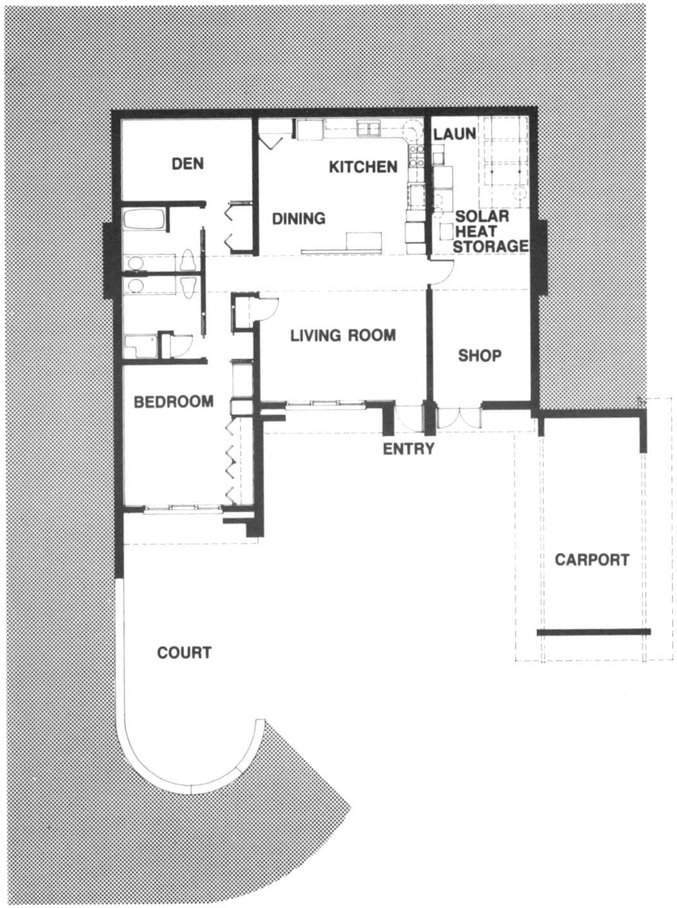
DEN
KITCHEN
LAUN
DINING
SOLAR
HEAT
STORAGE
LIVING ROOM
SHOP
BEDROOM
ENTRY
CARPORT
COURT
0 2 4 8 16

Floor Plan

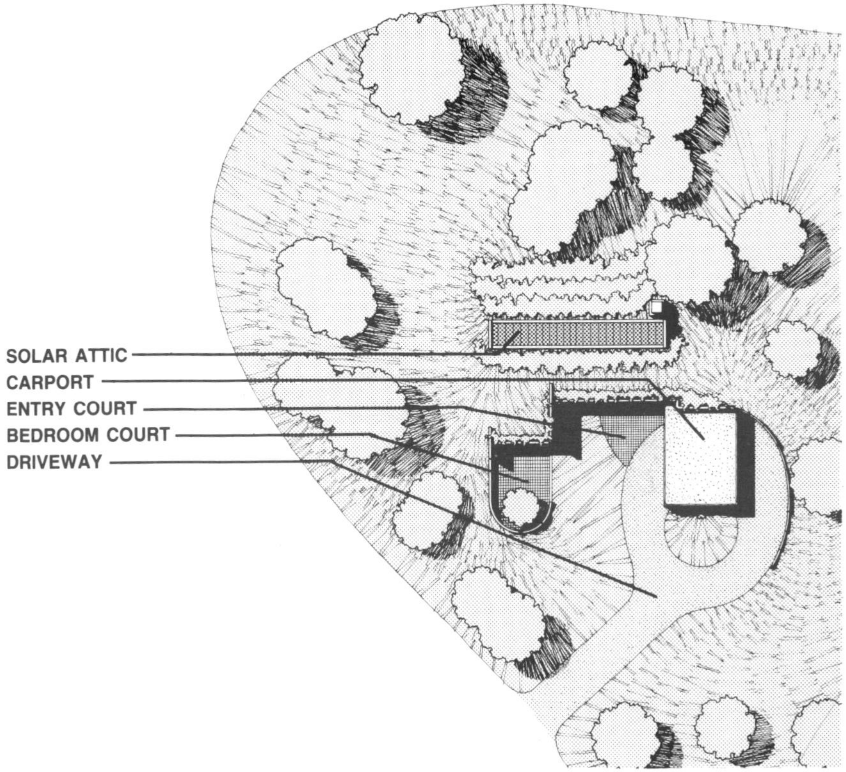
SOLAR ATTIC
CARPORT
ENTRY COURT
BEDROOM COURT
DRIVEWAY
N
0 10 20 40

Site Plan (aerial view)

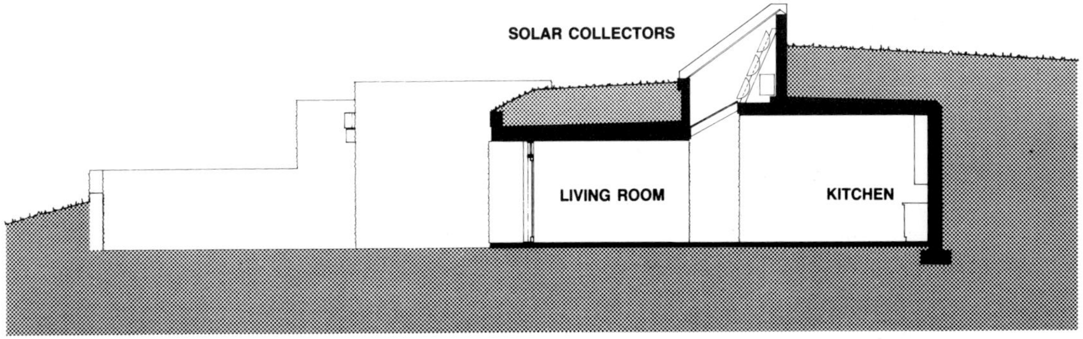
SOLAR COLLECTORS
LIVING ROOM
KITCHEN
0 2 4 8 16

Section

LOCATION:	Asheville, North Carolina
ARCHITECT:	Richard F. Webster
STRUCTURAL ENGINEER:	Bernard Feinberg
MECHANICAL ENGINEER:	Reece, Noland & McElrath
CONTRACTOR:	Richard Kennedy
CONSTRUCTED:	May 1978
PHOTOGRAPHY:	Richard F. Webster
GROSS AREA:	1,920 sq. ft. (173 ca)
STRUCTURE:	Reinforced concrete block walls, precast concrete roof, concrete slab-on-grade floor
EARTH COVER:	80% on roof at 30-45 in. (76-114 cm) 75% on walls
INSULATION:	Roof—3 in. (8 cm) rigid insulation Walls—2 in. (5 cm) rigid insulation
WATERPROOFING:	Butyl rubber membrane
HEATING DEGREE DAYS:	4,200
HEATING SYSTEM:	Solar attic, air-to-air heat pump, wood stove
COOLING SYSTEM:	Earth-cooled intake vent fan, air-to-air heat pump

Clark House

The Norman Clark home, like the Sticks & Stones house (pages 74-77), is an excellent illustration of how an earth sheltered house—even one with strikingly different design—can fit into an urban setting. The jutting projections on the house—a configuration that might well appear obtrusive in a conventional aboveground house—actually help the Clark home blend in with other houses in the neighborhood by making it appear less "underground." The hilly environment and a wealth of shrubs, trees, ivy, and other plantings surrounding the house also contribute to this sense of harmony with the environment.

Interestingly, it is the solar collectors, mounted on the concrete projection, that give the house its unique form. This shape evolved from architect Clark's desire to take advantage

of solar gain in a house that faces east—unlike most passive solar houses, which are oriented south. In fact, one of the main benefits of houses featuring an earth sheltered atrium design is that they do not require any specific orientation. In designing his house, Clark worked creatively with the opposing forces of the slope—which ruled out a south-facing house—and the need for a south orientation for the collectors.

The atrium design is a most appropriate plan for a house located in a rather dense suburban neighborhood such as this one. As with most such designs, the main living spaces—bedrooms, kitchen, dining and living rooms—are arranged around a central atrium that provides them with sunlight. The interior of the atrium, adorned with numerous plants and

shrubs, is an inviting family gathering place; in winter, it is converted to a greenhouse by stretching a polyethylene cover across the top of the space. The atrium/greenhouse also acts as a solar collector in winter, as it absorbs and stores heat. As shown in the floor plans, the kitchen and dining areas provide views of both the private, inner courtyard and, through exterior windows, the outdoors.

The house uses an active domestic hot water system, a 30-gallon (114-l.) electric water heater, and a fireplace as a backup. Additional solar energy obtained via the 400 square feet (36 ca) of rooftop solar collectors is stored in the 8-inch-thick, 8-foot-high (20-cm-thick, 2.4-m-high) walls and in the concrete slab floor.

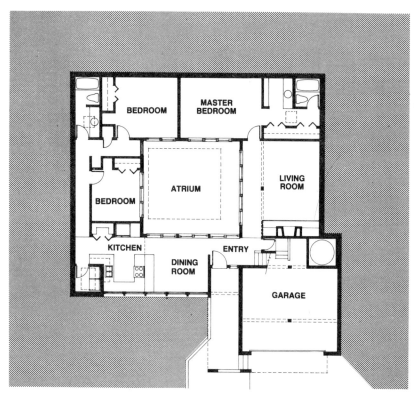

Floor Plan

0 2 4 8 16

SKYLIGHTS
ATRIUM
SOLAR
COLLECTORS

Site Plan (aerial view)

N→ 0 5 10 20

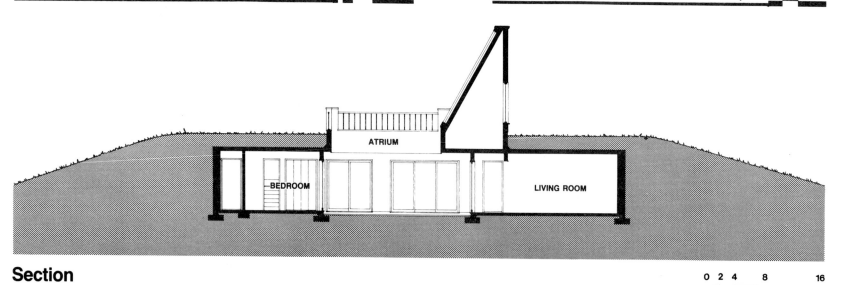

Section

0 2 4 8 16

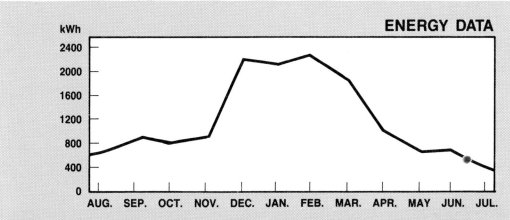

ENERGY DATA

kWh

| | AUG. | SEP. | OCT. | NOV. | DEC. | JAN. | FEB. | MAR. | APR. | MAY | JUN. | JUL. |

Total energy use for one year equals 18,515 KWh (this includes an estimated 3,600 KWh obtained from burning wood for space heating).

Norm Clark estimates that he uses 86-92% less energy for space heating than the average Oregon household based on data supplied by ODOE).

LOCATION:	Portland, Oregon
ARCHITECT:	Norm Clark
CONTRACTOR:	Norm Clark and Haldors Construction Company
CONSTRUCTED:	June 1977
GROSS AREA:	1,800 sq. ft. (162 ca) heated 500 sq. ft. (45 ca) unheated
STRUCTURE:	Cast-in-place reinforced concrete walls, wood beam and decking roof, concrete slab-on-grade floor
EARTH COVER:	80% on roof at 24 in. (61 cm) 88% on walls
INSULATION:	Roof—3 in. (8 cm) rigid insulation Walls—1 in. (2.5 cm) rigid insulation
WATERPROOFING:	Roof—five-ply hot mopped fiberglass Walls—asphalt coating
HEATING DEGREE DAYS:	4,700
HEATING SYSTEM:	Passive solar, wood, electric backup
COOLING SYSTEM:	Natural ventilation

Hadley House

To achieve his goal of using energy-saving techniques appropriately, the designer of this house integrated three major approaches to home energy conservation: earth sheltering, super insulation, and passive solar design. Through the use of these features, the architect has very successfully met the owner's requirements for brightly lit, aesthetically pleasing interior spaces; a high degree of energy efficiency; and a moderate construction budget.

The heavily wooded site for the house slopes rather steeply to the south. To increase usable outdoor space and provide adequate sunlight for passive heating, the designer and owner decided to detach the house from the garage and move it up the hill. On the north side, the house is set fully into the hill, with the earth sloping down on the west and east sides to meet grade on the south. By eliminating the traditional retaining walls, the architect kept construction costs down while only minimally reducing the thermal performance

of the house. The outdoor space to the south of the living spaces is designed so that a greenhouse and screened patio can be added.

The conceptual key to this south-facing home is the coupling of earth sheltering with passive solar direct gain. The sloping roof and large clerestory windows in every room maximize the penetration of sunlight to the house. The dark, insulated floor mass, together with the concrete block walls required to support the earth loads, serve as the storage medium for solar radiation. All low-mass wall surfaces are painted white to reflect and diffuse radiation so that it reaches all the dark high-mass interior surfaces.

Unique among the houses featured in this book, the Hadley house has no earth cover on the roof. Instead, the architect used large amounts of fiberglass insulation on the roof as well as throughout the house. Building the roof structure of 20-inch (50-cm) truss joists that span the entire width of the house left space for substantial amounts of insulation. The

designer estimates he saved the owner approximately $9,000-$10,000 in construction costs by using these "super insulation" techniques. Thermal performance data on the house have indicated that although the surface area of the roof accounts for 50 percent of the total exterior surface area of the house, it accounts for only 7 percent of the heat loss from the structure.

The primary heating system is electric forced-air, with supply ducts located below the floor slab. The air-return duct, which runs the length of the peak of the roof, collects and redistributes the warm air. To redistribute warm air from either the fireplace/wood stove or from solar heat gain, the high air return can be used automatically with the furnace blower.

More information about prototype house plans based on this house design can be obtained from: Tom Ellison, Ellison Design and Construction, 2001 University Avenue Southeast, Minneapolis, Minnesota 55414.

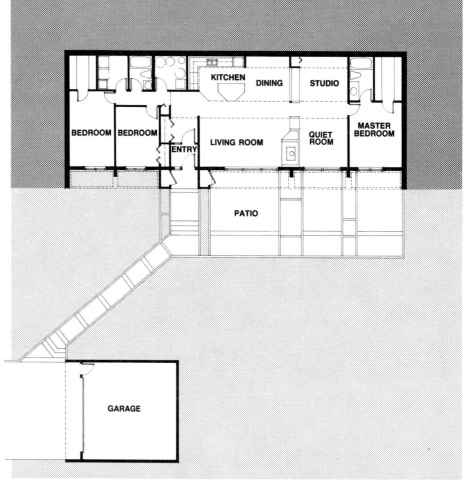

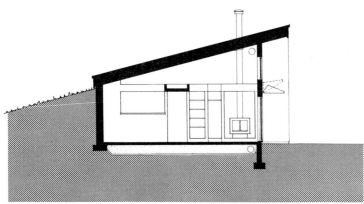

Floor Plan

KITCHEN	DINING	STUDIO

BEDROOM BEDROOM

ENTRY LIVING ROOM QUIET ROOM MASTER BEDROOM

PATIO

GARAGE

0 2 4 8 16

Section

0 2 4 8 16

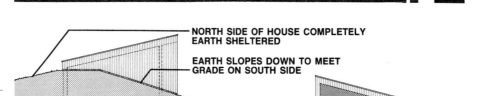

NORTH SIDE OF HOUSE COMPLETELY EARTH SHELTERED

EARTH SLOPES DOWN TO MEET GRADE ON SOUTH SIDE

Elevation

0 2 4 8 16

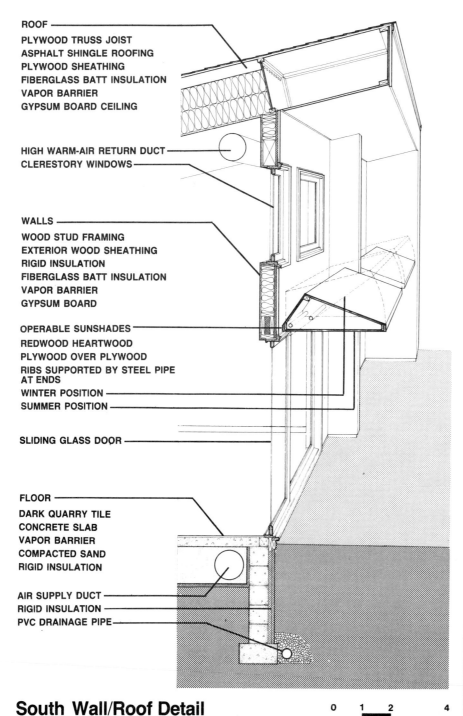

ROOF
PLYWOOD TRUSS JOIST
ASPHALT SHINGLE ROOFING
PLYWOOD SHEATHING
FIBERGLASS BATT INSULATION
VAPOR BARRIER
GYPSUM BOARD CEILING

HIGH WARM-AIR RETURN DUCT
CLERESTORY WINDOWS

WALLS
WOOD STUD FRAMING
EXTERIOR WOOD SHEATHING
RIGID INSULATION
FIBERGLASS BATT INSULATION
VAPOR BARRIER
GYPSUM BOARD

OPERABLE SUNSHADES
REDWOOD HEARTWOOD
PLYWOOD OVER PLYWOOD
RIBS SUPPORTED BY STEEL PIPE
AT ENDS
WINTER POSITION
SUMMER POSITION

SLIDING GLASS DOOR

FLOOR
DARK QUARRY TILE
CONCRETE SLAB
VAPOR BARRIER
COMPACTED SAND
RIGID INSULATION

AIR SUPPLY DUCT
RIGID INSULATION
PVC DRAINAGE PIPE

South Wall/Roof Detail

0 1 2   4

LOCATION:	Minneapolis, Minnesota
ARCHITECT:	Tom Ellison
CONTRACTOR:	Ellison Design & Construction
CONSTRUCTED:	1979
PHOTOGRAPHY:	John Fulker, Tom Ellison
GROSS AREA:	1,950 sq. ft. (176 ca)
STRUCTURE:	Reinforced concrete block walls, wood truss roof, concrete slab-on-grade floor
EARTH COVER:	50% on walls
INSULATION:	Roof—18 in. (46 cm) fiberglass Walls—4½ in. (11 cm) rigid insulation, tapering to 1½ in. (4 cm)
WATERPROOFING:	Bituthene on walls, asphalt shingles on roof
HEATING DEGREE DAYS:	8,300
HEATING SYSTEM:	Passive solar, wood stove, electric forced-air furnace
COOLING SYSTEM:	Natural ventilation

Feuille House

This house in The Sea Ranch, California, is a good example of carefully combining earth sheltering, design details, and landscaping to produce a house that is extremely compatible with the natural environment. The owner requested a simple, energy-conserving building that could easily withstand the harsh rains and strong winds typical of the coastal climate, while permitting a good view of the Pacific Ocean to the south. Benched into a south-facing, gently sloping hillside, the house has a sod roof planted with natural grasses and ice plant that help integrate the house with the site. By virtue of its strong, simple lines, the house takes its place as an integral part of the landscape, rather than overpowering it. Redwood siding and terra-cotta floor tiles reflect the colors of the surrounding terrain.

The south orientation of the house permits it to benefit from direct passive solar gain while offering its occupants an expansive view of the rugged northern California shoreline. Passive gain from the vertical, south-facing windows is stored in the concrete block retaining walls and in the tile floors on concrete slab. At the request of the owners, all the double-glazed glass was tinted to minimize glare. Three skylights provide additional spot lighting at the rear of the house.

A projecting tower that houses a loft also contains operable clerestory windows that assist with ventilation and provide additional light. Naturally heated stratified air is moved by an automated fan from the tower to a sand heat storage bed below the tile floor.

An interesting design detail is the placement of thermosiphonic solar domestic water heating collectors at ground level. The collectors are integrated with the earth slope below the exterior, south-facing wood deck.

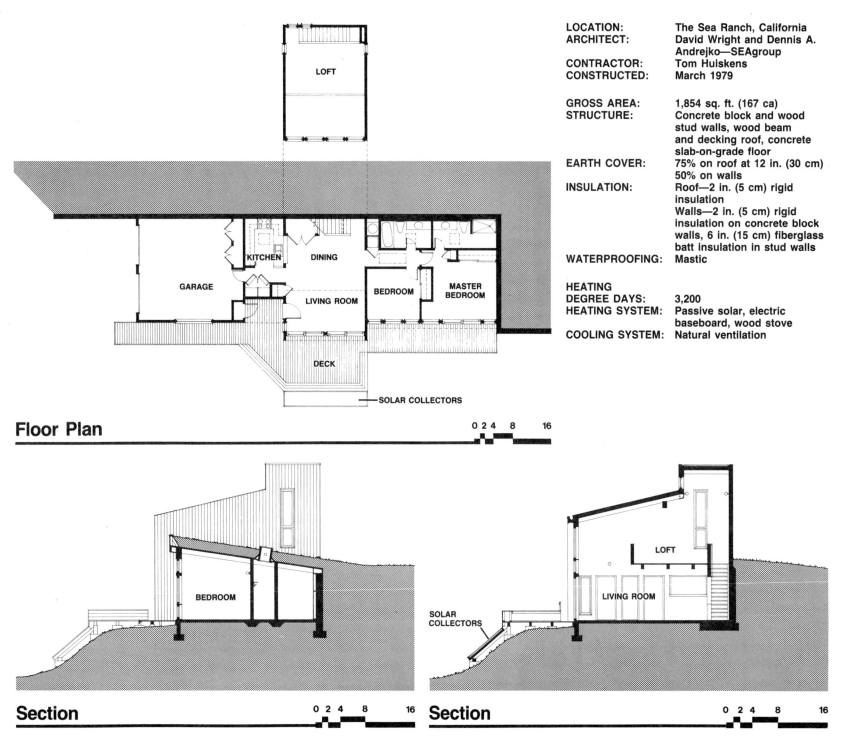

LOCATION:	The Sea Ranch, California
ARCHITECT:	David Wright and Dennis A. Andrejko—SEAgroup
CONTRACTOR:	Tom Huiskens
CONSTRUCTED:	March 1979
GROSS AREA:	1,854 sq. ft. (167 ca)
STRUCTURE:	Concrete block and wood stud walls, wood beam and decking roof, concrete slab-on-grade floor
EARTH COVER:	75% on roof at 12 in. (30 cm) 50% on walls
INSULATION:	Roof—2 in. (5 cm) rigid insulation Walls—2 in. (5 cm) rigid insulation on concrete block walls, 6 in. (15 cm) fiberglass batt insulation in stud walls
WATERPROOFING:	Mastic
HEATING DEGREE DAYS:	3,200
HEATING SYSTEM:	Passive solar, electric baseboard, wood stove
COOLING SYSTEM:	Natural ventilation

LOFT

KITCHEN DINING

GARAGE

LIVING ROOM

BEDROOM

MASTER BEDROOM

DECK

SOLAR COLLECTORS

Floor Plan

0 2 4 8 16

BEDROOM

Section

0 2 4 8 16

LOFT

LIVING ROOM

SOLAR COLLECTORS

Section

0 2 4 8 16

Sticks & Stones House

The Sticks & Stones house successfully carries through a scaled-down approach to earth sheltered design. The architects have designed both the interior living and storage spaces and the exterior courtyard areas thoughtfully and imaginatively, so that the overall house plan seems comfortably compact, rather than cramped. A high degree of natural sunlight admitted to the rooms through south-facing windows and two window courts, in combination with a relatively spacious living/dining/sun court area, contribute to a sense of airiness and openness.

This house also is a fine example of how earth sheltering can be used to make the most of a small site in a typical urban setting. The earth berm presents a low profile in scale with the lot, and provides ample yard space—space that would otherwise have been consumed by a

conventional, aboveground home. The careful landscaping, using trees that were growing on the site before construction as well as planter trees and flowers, helps the house blend into the neighborhood while preserving precious "green space" in an urban environment.

With earth covering 90 percent of the roof and 80 percent of the sides, this house may best be described as a penetrational, as opposed to elevational or atrium, design. This type of house plan is characterized by an extensive earth cover on the roof and by window openings on several sides of the house. The east window court in this house helps provide natural light to the two main bedrooms, which are set away from the primary living spaces.

A very effective design feature in the main living area is the raised wooden floor, which

subtly separates the living and dining areas in an otherwise open space. To enhance the feeling of spaciousness in this area, the precast ceiling is higher here than in the bedrooms. The sunken sun court provides sunlight and ventilation to the rooms and access to the roof.

Building heat is accumulated from several sources. The primary heating system—a gas-fired, forced-air furnace—is supplemented by passive solar heat gain through the double-glazed, south-facing glass and by a wood-burning stove in the raised living room space. Heat is stored and released from the thermal mass of the walls, roof, and floor.

The conventional roof of the carport provides storage space and presents a sloped surface for future installation of solar panels.

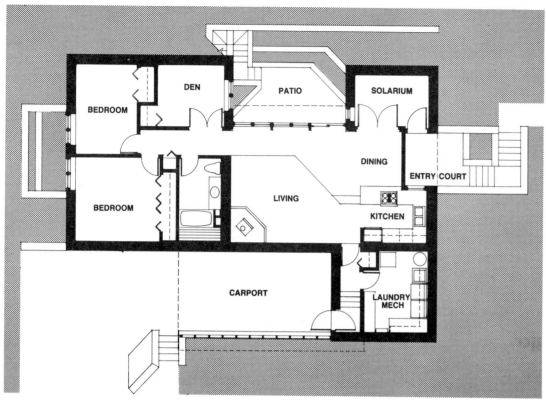

Floor Plan

0 2 4 8 16

Section

0 2 4 8 16

Site Plan (aerial view)

0 10 20

BEDROOM WINDOW COURT
DRIVEWAY
CARPORT
ENTRY COURT
SUNKEN PATIO

LOCATION:	**Minneapolis, Minnesota**
ARCHITECT:	**Froehle, Saphir, Joos—Sticks & Stones Design**
ENGINEER:	**Fowler, Hanley**
CONTRACTOR:	**Sticks & Stones Design**
CONSTRUCTED:	**1980**
PHOTOGRAPHY:	**Steven L. Bergerson**
GROSS AREA:	**1,350 sq. ft. (122 ca)**
STRUCTURE:	**Reinforced concrete block walls, precast concrete roof, concrete slab-on-grade floor**
EARTH COVER:	**90% on roof at 12-18 in. (30-46 cm)**
	80% on walls
INSULATION:	**Roof—6 in. (15 cm) rigid insulation**
	Walls—3 in. (7 cm) rigid insulation
WATERPROOFING:	**Bituthene**
HEATING DEGREE DAYS:	**8,000**
HEATING SYSTEM:	**Gas forced-air furnace, passive solar, wood stove**
COOLING SYSTEM:	**Natural ventilation**

Earthtech 5 and 6

These two houses are variations on the prototype Earthtech house design developed by architect Don Metz, a pioneer in earth sheltered construction. Both houses are characterized by simplicity and durability in design and construction—themes echoed in the choice of building materials. Inside Earthtech 5, for example, the heavy timber roof system, large masonry arch over the wood stove, and dark floor tiles link the house to its wooded site and help create a warm, comfortable interior environment. The cedar shiplap siding that has been used on the exterior surfaces has been bleached light gray and left to weather naturally.

Exposure of both the west and south sides of these two houses is a rather unusual design feature. Although these designs may not take as much advantage of passive solar opportunities as some other earth sheltered houses, they offer considerable natural lighting and a choice of views not generally available with the more common lineal arrangements of south-facing earth sheltered homes. The exposed west wall also provides a means of easy emergency exit from the bedrooms.

Like most earth-integrated structures, these homes were designed with passive solar gain in mind. Heat obtained from the sun coming through the south and west glazed walls is absorbed and dispensed by the built-in thermal mass of the concrete walls, the floor, and the roof system.

The floor system is another feature associated with these Earthtech designs. A double slab air plenum, the floor consists of a concrete slab supported by concrete blocks spaced at intervals on top of a slab on grade. Fans force warm air into the spaces between the blocks, distributing it evenly, so that the floor is always warmer than a typical slab-on-grade floor. As a result of this floor system with its large, warm thermal mass, temperature changes within the house are not felt to the same extent that they would be in houses with less thermal mass.

More information about the prototype Earthtech plans may be obtained by writing to: Don Metz, Earthtech, Post Office Box 52, Lyme, New Hampshire 03768.

Floor Plan

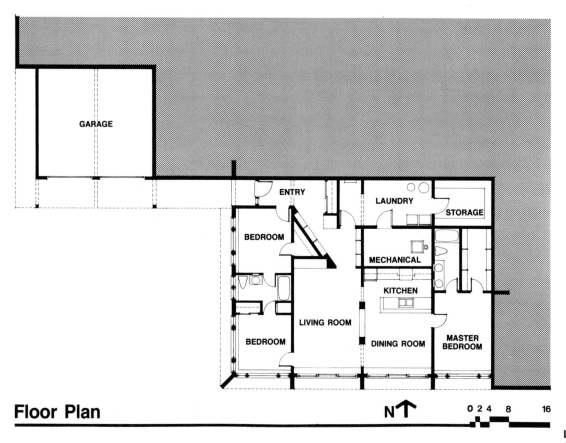

GARAGE

ENTRY

LAUNDRY

STORAGE

BEDROOM

MECHANICAL

KITCHEN

LIVING ROOM

BEDROOM

DINING ROOM

MASTER BEDROOM

N↑ 0 2 4 8 16

Section

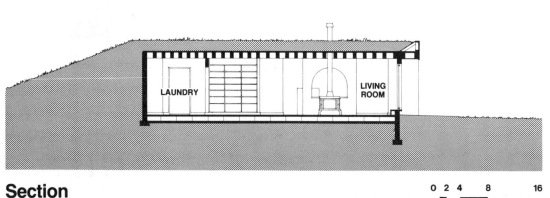

LAUNDRY

LIVING ROOM

0 2 4 8 16

LOCATION:	**Lyme, New Hampshire**
ARCHITECT:	**Don Metz**
CONTRACTOR:	**Wayne Pike**
CONSTRUCTED:	**1979**
PHOTOGRAPHY:	**©1979 Robert Perron; Don Metz**
GROSS AREA:	**2,000 sq. ft. (180 ca)**
STRUCTURE:	**Reinforced concrete walls, heavy timber beam and tongue and groove decking roof, double-slab concrete floor**
EARTH COVER:	**100% on roof at 9 in. (23 cm) 50% on walls**
INSULATION:	**Roof—4 in. (10 cm) rigid insulation Walls—2 in. (5 cm) rigid insulation**
WATERPROOFING:	**Celotex IRA**
HEATING DEGREE DAYS:	**7,600**
HEATING SYSTEM:	**Oil-fired hydronic system, wood stove**
COOLING SYSTEM:	**Natural ventilation**

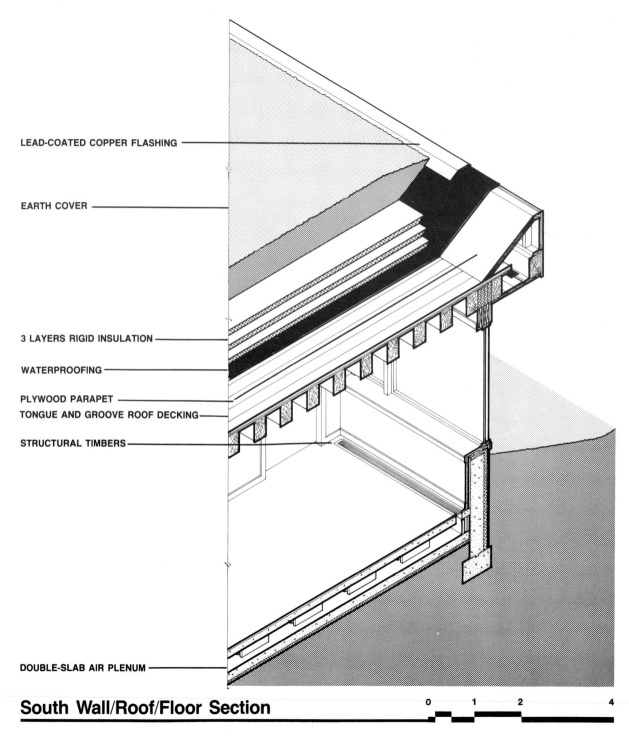

LEAD-COATED COPPER FLASHING

EARTH COVER

3 LAYERS RIGID INSULATION

WATERPROOFING

PLYWOOD PARAPET

TONGUE AND GROOVE ROOF DECKING

STRUCTURAL TIMBERS

DOUBLE-SLAB AIR PLENUM

South Wall/Roof/Floor Section

0   1   2   4

SunEarth House

The SunEarth house in Colorado serves as an excellent example of how passive solar technology can be combined with a basic earth sheltered house design to yield quite significant energy savings over conventional homes. The low energy consumption of this house results both from the passive solar features incorporated into the house and the owners' knowledge of how to use these features to obtain the best energy performance.

With berms on the roof and on the east, west, and north sides, the SunEarth house has 300 square feet (28 ca) of floor-to-ceiling south-facing windows. Fifty-four 55-gallon (208-l.) water-filled black drums are stacked against these windows. Air heated by the drums rises into the ceiling vents above the drums, then flows into an open plenum between the ceiling

and the roof. Vents from this plenum provide a thermosiphonic flow of warm air around the inside of the house. Although most of the solar energy collected through the south windows is stored in the drums, the mass of the slab floor and the walls provide some additional storage capacity.

The double glazing on the south windows is protected by a Beadwall movable insulation system consisting of a wall glazed with parallel panes of plate glass spaced 5 inches (12.5 cm) apart. In cloudy weather and at night, tiny polystyrene beads are automatically blown in between the panes.

Six triple-glazed sunscoop skylights extending up through the berms add to the natural lighting provided by the south windows. Painted a reflective white on the inside, these

skylights are tilted to the south to increase light gain in winter and minimize solar gain in summer.

Water preheating is handled by two 30-gallon (114-l.) black steel domestic hot water tanks, placed in slanted enclosures on the exposed south wall, that collect heat from the sun in summer and winter.

Carefully monitored for energy performance over the past two years, SunEarth house uses energy from the sun to handle all the space heating and 20 percent of the domestic hot water needs for the house. In addition to the obvious economic benefits, the owners cite the openness of the house plan, low maintenance, and quietness as other positive aspects of their earth sheltered home.

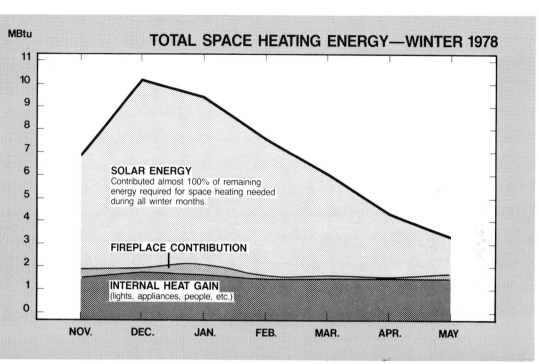

TOTAL SPACE HEATING ENERGY—WINTER 1978

MBtu

SOLAR ENERGY
Contributed almost 100% of remaining energy required for space heating needed during all winter months.

FIREPLACE CONTRIBUTION

INTERNAL HEAT GAIN
(lights, appliances, people, etc.)

NOV. DEC. JAN. FEB. MAR. APR. MAY

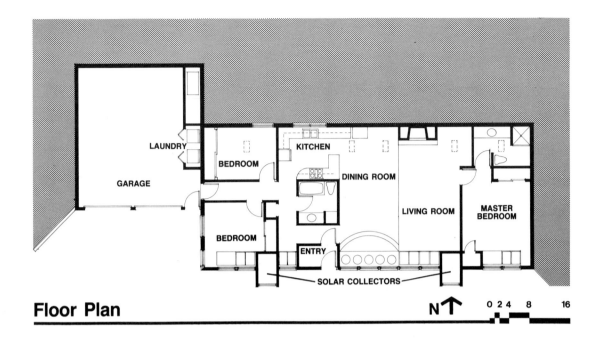

Floor Plan

LAUNDRY

GARAGE

BEDROOM

KITCHEN

DINING ROOM

BEDROOM

ENTRY

LIVING ROOM

MASTER BEDROOM

SOLAR COLLECTORS

N↑   0 2 4   8   16

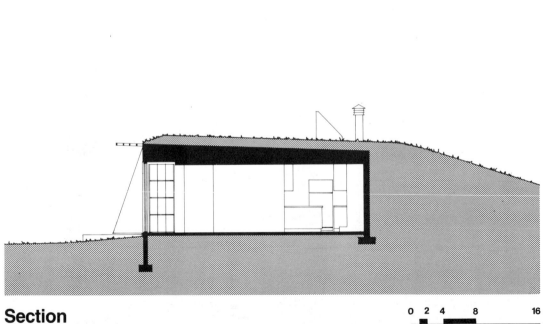

Section

0 2 4   8   16

LOCATION:	**Longmont, Colorado**
DESIGNER:	**Paul Shippee—Colorado Sunworks**
STRUCTURAL ENGINEER:	**Johnson-Voiland-Archulela, Inc.**
CONTRACTOR:	**Colorado Sunworks**
CONSTRUCTED:	**1978**
GROSS AREA:	**1,800 sq. ft. (162 ca)**
STRUCTURE:	**Cast-in-place concrete walls, concrete deck on steel joist roof, concrete slab-on-grade floor**
EARTH COVER:	**100% on roof at 12 in. (30 cm) 75% on walls**
INSULATION:	**Roof—2 in. (5 cm) rigid insulation, 9 in. (23 cm) fiberglass batt insulation Walls—4 in. (10 cm) rigid insulation**
WATERPROOFING:	**Three-ply tar and felt**
HEATING DEGREE DAYS:	**6,200**
HEATING SYSTEM:	**Passive solar, hot water baseboard, heat-circulating fireplace**
COOLING SYSTEM:	**Three turbine ventilators**

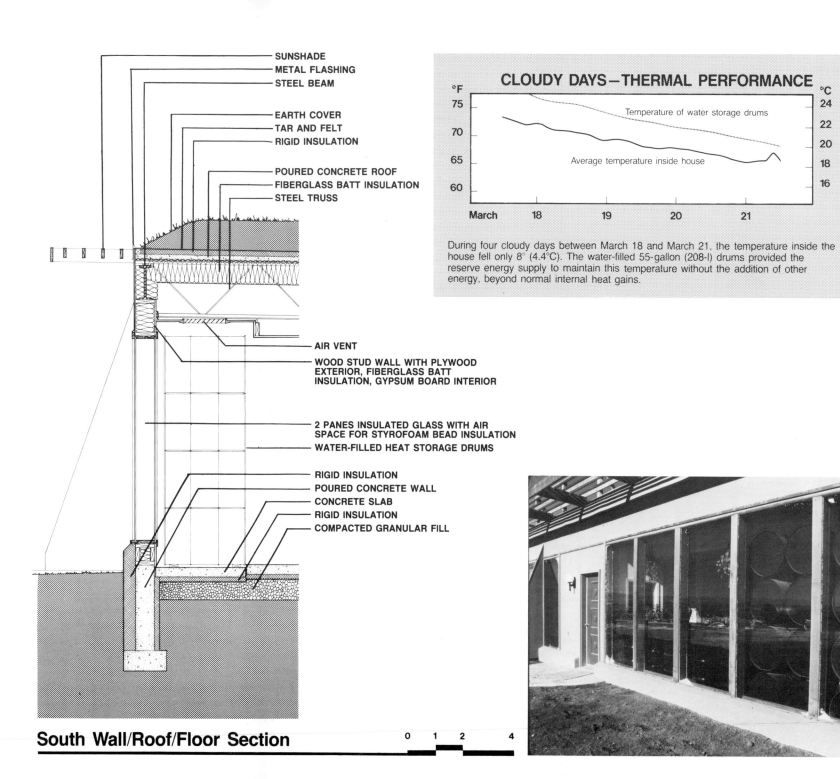

SUNSHADE
METAL FLASHING
STEEL BEAM

EARTH COVER
TAR AND FELT
RIGID INSULATION

POURED CONCRETE ROOF
FIBERGLASS BATT INSULATION
STEEL TRUSS

CLOUDY DAYS—THERMAL PERFORMANCE

Temperature of water storage drums

Average temperature inside house

°F: 75, 70, 65, 60
°C: 24, 22, 20, 18, 16

March 18 19 20 21

During four cloudy days between March 18 and March 21, the temperature inside the house fell only 8° (4.4°C). The water-filled 55-gallon (208-l) drums provided the reserve energy supply to maintain this temperature without the addition of other energy, beyond normal internal heat gains.

AIR VENT
WOOD STUD WALL WITH PLYWOOD EXTERIOR, FIBERGLASS BATT INSULATION, GYPSUM BOARD INTERIOR

2 PANES INSULATED GLASS WITH AIR SPACE FOR STYROFOAM BEAD INSULATION
WATER-FILLED HEAT STORAGE DRUMS

RIGID INSULATION
POURED CONCRETE WALL
CONCRETE SLAB
RIGID INSULATION
COMPACTED GRANULAR FILL

South Wall/Roof/Floor Section

0 1 2 4

Boothe House

Architects in southern Texas, where this home is located, have a twofold problem with regard to energy use. The greatest demands on climate control systems occur during summer, when high temperatures average in the upper nineties and often exceed 100°F (38°C) for days. On the other hand, winter brings sporadic below-freezing temperatures accompanied by extremely changeable weather conditions (e.g., temperatures have been known to plummet from 76°F (25°C) to 24°F (–4°C) within hours when a "blue norther" blows in). The period of highest heating demand lasts from mid-December to mid-March.

The architect has combined earth sheltering with passive solar features to deal efficiently with both heating and cooling needs of houses in this climate. And given the rising cost of fossil fuels, earth sheltering has become an increasingly popular option in this region of the country.

The primary design requirements requested by the retired owner of this home were reduced utility costs (including decreased dependence on energy derived from fossil fuel), ease of maintenance, simplicity of construction, and security. Attractive landscaping, quietness, and "ecological aspects" are additional benefits the owner attributes to her earth sheltered home.

The house, built into a southeast-facing hillside, takes advantage of the view, orientation, and prevailing breezes. In winter, the house is partially heated by warm air collected in the greenhouse. Active solar collectors are used to heat the domestic water, which comes from a deep well on the 15-acre (6-ha) property.

The centrally located "wind tower" presents a unique solution to the common problem of providing adequate cross-ventilation in an earth

sheltered house. The wind tower, which can be opened and closed from inside the house, draws the prevailing southwest winds through the house in summer, offering welcome cross-ventilation and cooling. Shaded roof overhangs help keep the sun away from the house in summer. A small, packaged air conditioning unit acts as a backup cooling system.

During the summer of 1980, which broke records for temperature highs, number of consecutive days over 100°F (38°C), and lack of rainfall, the house turned in an extremely good energy performance. Even though landscaping had scarcely begun and the roof was covered mainly with a sparse crop of weeds, the house consumed only about 40 percent of the energy required by similar-sized homes in the area. Once the landscaping has become established, even greater energy efficiency is expected.

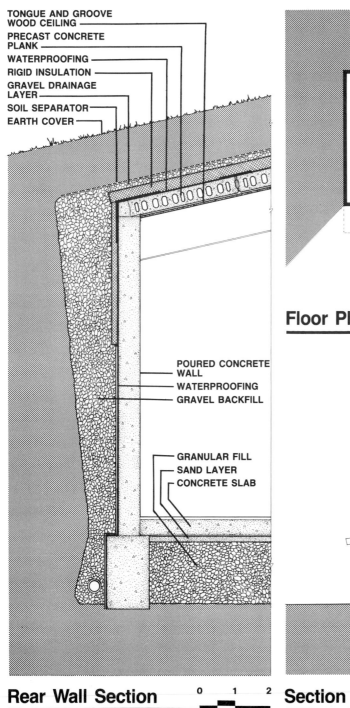

TONGUE AND GROOVE
WOOD CEILING
PRECAST CONCRETE
PLANK
WATERPROOFING
RIGID INSULATION
GRAVEL DRAINAGE
LAYER
SOIL SEPARATOR
EARTH COVER

POURED CONCRETE
WALL
WATERPROOFING
GRAVEL BACKFILL

GRANULAR FILL
SAND LAYER
CONCRETE SLAB

Rear Wall Section   0   1   2

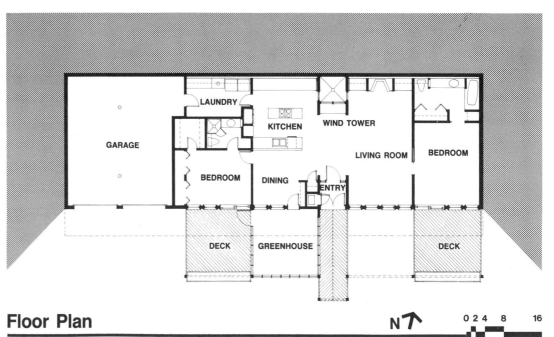

GARAGE

LAUNDRY

KITCHEN

WIND TOWER

BEDROOM

DINING

ENTRY

LIVING ROOM

BEDROOM

DECK   GREENHOUSE   DECK

Floor Plan   N↑   0 2 4   8   16

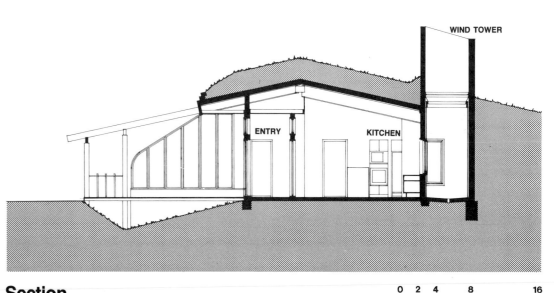

WIND TOWER

ENTRY   KITCHEN

Section   0 2 4   8   16

89

LOCATION:	Fort Worth, Texas
ARCHITECT:	Ray W. Boothe—Boothe & Associates
STRUCTURAL ENGINEER:	William D. Walker
MECHANICAL ENGINEER:	James McClure
CONTRACTOR:	Terra Systems
CONSTRUCTED:	May 1980
PHOTOGRAPHY:	Kipp Baker

GROSS AREA:	2,784 sq. ft. (251 ca)
STRUCTURE:	Cast-in-place concrete walls, precast planks on structural steel frame roof, concrete slab-on-grade floor
EARTH COVER:	100% on roof at 24 in. (61 cm) 66% on walls
INSULATION:	Roof—4 in. (10 cm) rigid insulation Walls—2 in. (5 cm) rigid insulation
WATERPROOFING:	Bentonite

HEATING DEGREE DAYS:	3,041
COOLING DEGREE DAYS:	2,965
HEATING SYSTEM:	Electric forced-air system, solar greenhouse, heat-circulating fireplace
COOLING SYSTEM:	Wind tower, natural ventilation

Wells House/Office

"When you have a stunning pond and forest view to the east you build facing east and find some other way to admit solar radiation." So says architect Malcolm Wells, a pioneer in earth sheltered house design. Wells' combination home/office on Cape Cod is one room wide and 110 feet (33 m) long, with its long axis running due north-south. Along the full length of this axis runs a centrally located, triple-glazed skylight, which admits substantial natural light and solar heat to all the rooms in the house. Snap-on canvas shades shield the skylight during the summer months to avoid overheating.

Wells has taken advantage of the aesthetic benefits of earth sheltering to create a house that blends extremely well with the surrounding woods. Moreover, the varying heights and depths of the earth berms—allowing for entrances, windows, and vents—help soften and break up the long facade of the home. By contrast, an aboveground building designed in a similar style—i.e., one with a long, relatively narrow rectangular shape—would have had a harsh, stark appearance in this landscape.

Mrs. Wells, a landscape designer, has used mostly native trees, shrubs, and ground cover from the site to good effect. The landscaping on the roof is more noticeable on the Wells house than is the case with most other earth sheltered houses because Wells has avoided using a curb on the edge of the roof. Tapering the earth and vegetation to the eave edge not only eliminated the need for the traditional curbing, but also helps integrate the rooftop landscaping with the overall house design.

Wells says he sometimes jokes to visitors that the smooth, dark floors are slabs of imported Italian leather. This unusual effect was, in fact, rather simply achieved by coating the exposed concrete floor with a mixture of dark wood stain and urethane sealer. Wells notes that the floor is easy to clean, absorbs sunlight well, and is rough enough to eliminate the problems with slipperiness sometimes associated with concrete floor slabs. The heating pipes located between the floor slabs also keep the concrete floor warm enough to walk on in bare feet.

Ducts pull the hottest air down from the top of the skylight, push it through air pipes between the floor slabs, and return it at a temperature of approximately 70°-75°F (21°-24°C).

More information about similar earth sheltered house plans is available in *Underground Designs* ($6.00 postpaid) and *Underground Plans Book I* ($13.00 postpaid), which may be ordered from Malcolm Wells, Post Office Box 1149, Brewster, Massachusetts 02631.

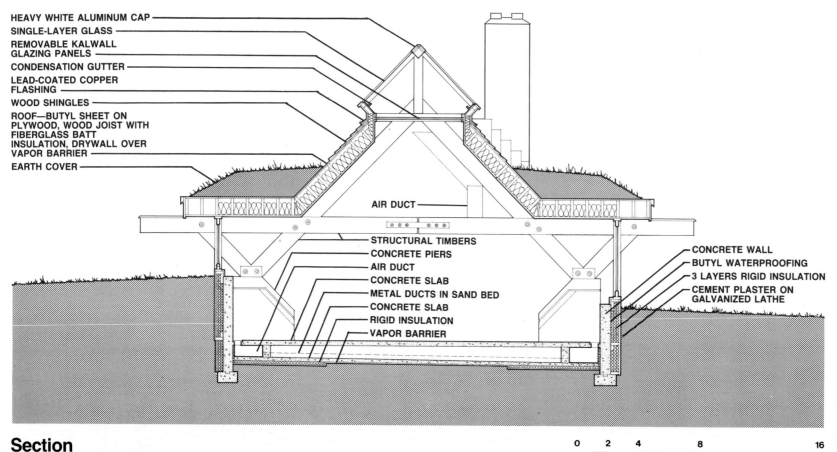

HEAVY WHITE ALUMINUM CAP

SINGLE-LAYER GLASS

REMOVABLE KALWALL
GLAZING PANELS

CONDENSATION GUTTER

LEAD-COATED COPPER
FLASHING

WOOD SHINGLES

ROOF—BUTYL SHEET ON
PLYWOOD, WOOD JOIST WITH
FIBERGLASS BATT
INSULATION, DRYWALL OVER
VAPOR BARRIER

EARTH COVER

AIR DUCT

STRUCTURAL TIMBERS

CONCRETE PIERS

AIR DUCT

CONCRETE SLAB

METAL DUCTS IN SAND BED

CONCRETE SLAB

RIGID INSULATION

VAPOR BARRIER

CONCRETE WALL

BUTYL WATERPROOFING

3 LAYERS RIGID INSULATION

CEMENT PLASTER ON
GALVANIZED LATHE

Section

0   2   4   8   16

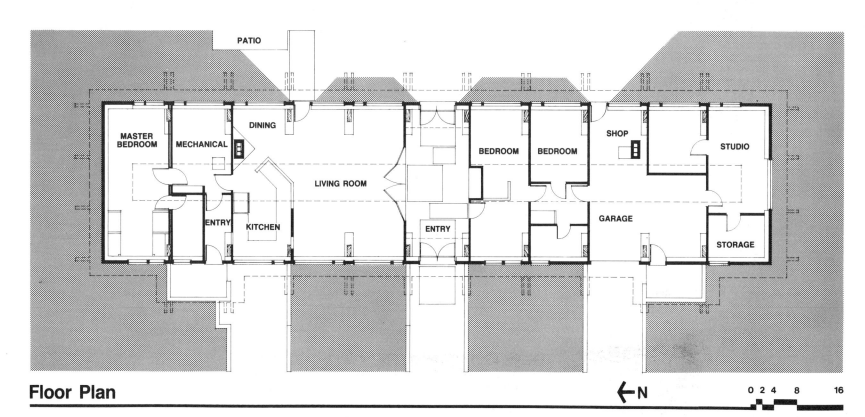

Floor Plan

PATIO

MASTER BEDROOM

MECHANICAL

DINING

LIVING ROOM

ENTRY

KITCHEN

ENTRY

BEDROOM

BEDROOM

SHOP

STUDIO

GARAGE

STORAGE

←N

0 2 4 8 16

LOCATION:	Brewster, Massachusetts
ARCHITECT:	Malcolm Wells
SOLAR CONSULTANT:	Robert O. Smith
CONTRACTOR:	Sol Source Builders
CONSTRUCTED:	May 1980
GROSS AREA:	2,600 sq. ft. (234 ca)
STRUCTURE:	Poured concrete walls, wood beam and joist roof, double-slab concrete floor
EARTH COVER:	45% on roof at 18 in. (46 cm) 40% on walls
INSULATION:	Roof—9 in. (23 cm) fiberglass batt insulation Walls—6 in. (15 cm) rigid insulation
WATERPROOFING:	E.P.D.M. rubber membrane on roof and walls
HEATING DEGREE DAYS:	5,621
HEATING SYSTEM:	Oil-fired hot water, passive solar
COOLING SYSTEM:	Natural ventilation

Archititerra Houses

Among the houses included in this book, the Architerra homes are unique in two ways: they are the only examples of homes from outside the United States, and they were specifically designed to be built on hillsides. And, like the Seward town houses, these homes are unusual in that they are among the few multiresidential earth sheltered developments built to date. Several Architerra complexes have been completed in France and Spain, and the company is seeking developers in the United States.

These developments serve as an excellent example of how earth sheltering can turn a site considered unsuitable for conventional construction to aesthetic—and economic—advantage. The Architerra construction system is extremely well suited for developments on steep slopes because of its flexibility and ability to stabilize hillside conditions.

By virtue of the stepped design of the Architerra complexes and the curved glass front walls of the units, all the residents of the development enjoy a 180-degree view of the surrounding landscape. These south-facing glass walls provide natural light and view and passive solar gain and help the units appear naturally integrated with the curves of the hillside.

Stacking the units stair fashion in slots cut into the hillside imparts a sense of privacy to each unit—because of the unobstructed view of the horizon—while simultaneously permitting a high unit density (e.g., an estimated ten units per acre in the recently completed Nice project). Surface preservation and use of land that would otherwise be considered unusable are obvious environmental benefits associated with this type of development.

The individual units are designed for flexibility in space arrangements. For example, the one- to four-bedroom units in the Nice

development vary from 1,000 to 4,000 square feet (90 to 360 ca) in size; room sizes vary depending on the owners' wishes. In most Architerra complexes, each unit has a private garden/courtyard.

The key to the Architerra system is the ability of the developers to mold the land and design through a patented construction technique called Reinforced Earth. First, the land is contoured into a series of terraces and excavated for the home sites. Then the Reinforced Earth technique—in which long metal strips attached to precast concrete walls are laid between successive layers of backfill—is used to stabilize the backfill (see photograph on page 101). The wall panels, which form the curved rear structural walls of the units, resist lateral earth pressures very efficiently.

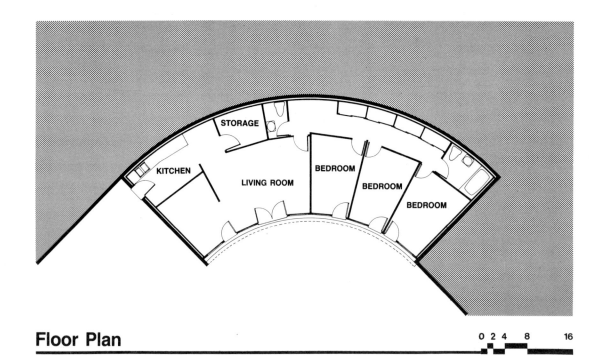

Floor Plan

STORAGE

KITCHEN

LIVING ROOM

BEDROOM

BEDROOM

BEDROOM

0 2 4 8 16

Section

LIVING ROOM

0 2 4 8 16

LOCATION:	Valbonne, France and Madrid, Spain
ARCHITECT:	Architerra, Inc.
DESIGNER:	Henri Vidal, Yves Bayard
CONTRACTOR:	M. Nicoletti (Valbonne), M. Santiago Muras Muerano (Madrid)
CONSTRUCTED:	1979
PHOTOGRAPHY:	The Washington Agency
GROSS AREA:	1,000-4,000 sq. ft. (90-360 ca)
STRUCTURE:	Precast concrete walls, cast-in-place concrete roof, concrete slab-on-grade floor
EARTH COVER:	100% on roof at 12 in. (30 cm) 50% on walls
INSULATION:	Roof—1 in. (2.5 cm) rigid insulation Walls—4 in. (10 cm) air space between precast walls and interior finish walls
WATERPROOFING:	Asphalt membrane and PVC
HEATING SYSTEM:	Electric baseboard
COOLING SYSTEM:	Natural ventilation

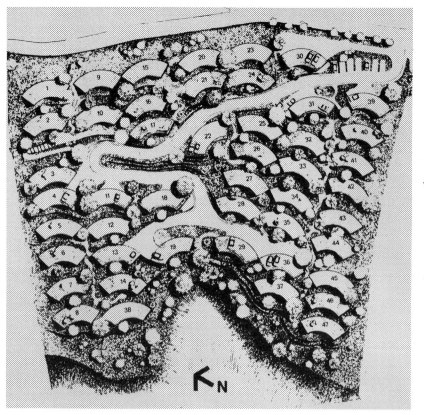

Site Plan (aerial view)

0 20   60   120

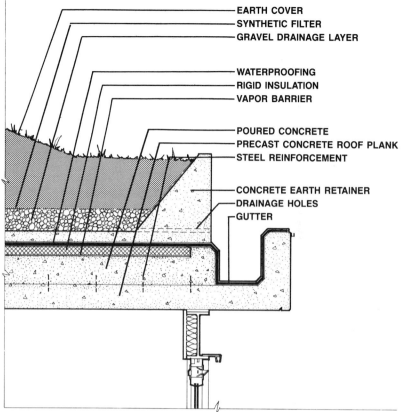

EARTH COVER
SYNTHETIC FILTER
GRAVEL DRAINAGE LAYER

WATERPROOFING
RIGID INSULATION
VAPOR BARRIER

POURED CONCRETE
PRECAST CONCRETE ROOF PLANK
STEEL REINFORCEMENT

CONCRETE EARTH RETAINER
DRAINAGE HOLES
GUTTER

Roof Parapet/Gutter Detail

0   1   2

Terra-Dome House

Undoubtedly, the most unusual feature of this Terra-Dome home is its modular concrete shell structure, using combinations of 24-foot-wide (7.3-m) modules. The patented dome forms used in constructing the house are common to all homes built by the Missouri-based Terra-Dome Corporation; this home is composed of six-and-a-half modules.

The entryway to the home, formed by half a module, is closed off from the rest of the house by a sliding glass door, which creates an air lock and uses the home's southern exposure to its fullest potential. The entry acts as a sunroom by collecting passive solar heat through the south-facing entrance windows and stores heat in the massive stone floors.

The primary heat source for the house, a wood-burning stove, is supplemented by the passive solar gain through the south-facing windows. Heat from the fireplace is radiated into a heat chamber behind it, from which the warm air is picked up and circulated through the vent pipes under the floor to all the rooms in the house.

One advantage to the dome-shaped modules is that the roofs can support heavy earth loads very efficiently—the builder claims these structures have twenty times the load-bearing capacity of flat-roofed earth shelters. Because of this increased load-bearing capacity, Terra-Dome homes normally are covered with 2 to 8 feet (.6 to 2.4 m) of earth, allowing the designers to gently grade soil over the structure following the natural curves of the earth.

Another advantage to this type of structure and building process is that the walls and roof are poured as one unit, leaving no cold joints at the roof lines. In combination, the standardized module forms and pouring system permit rapid—and, hence, less costly—construction.

Subdivisions, apartment complexes, and commercial space are currently under development using the Terra-Dome system.

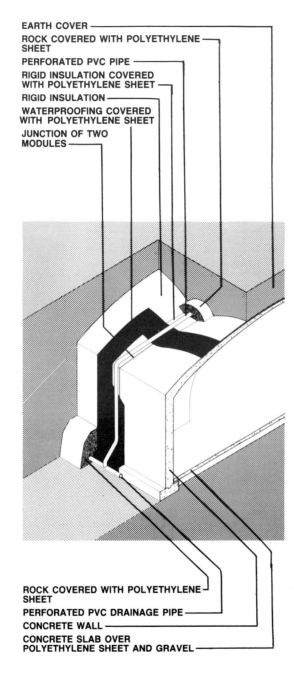

EARTH COVER

ROCK COVERED WITH POLYETHYLENE
SHEET

PERFORATED PVC PIPE

RIGID INSULATION COVERED
WITH POLYETHYLENE SHEET

RIGID INSULATION

WATERPROOFING COVERED
WITH POLYETHYLENE SHEET

JUNCTION OF TWO
MODULES

ROCK COVERED WITH POLYETHYLENE
SHEET

PERFORATED PVC DRAINAGE PIPE

CONCRETE WALL

CONCRETE SLAB OVER
POLYETHYLENE SHEET AND GRAVEL

Dome/Wall Section

0 2 4   8

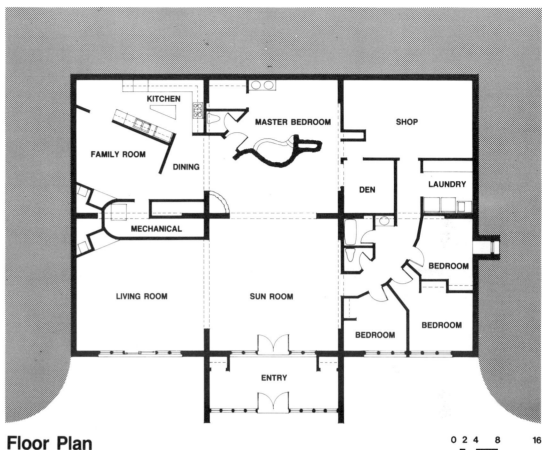

Floor Plan

0 2 4   8   16

KITCHEN

MASTER BEDROOM

SHOP

FAMILY ROOM

DINING

DEN

LAUNDRY

MECHANICAL

BEDROOM

LIVING ROOM

SUN ROOM

BEDROOM

BEDROOM

BEDROOM

ENTRY

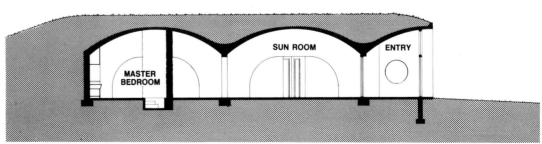

Section

0 2 4   8   16

MASTER
BEDROOM

SUN ROOM

ENTRY

LOCATION:	Independence, Missouri
ARCHITECT:	Terra-Dome Corporation
CONTRACTOR:	Terra-Dome Corporation
CONSTRUCTED:	December 1979
PHOTOGRAPHY:	Holiday Productions

GROSS AREA:	3,900 sq. ft. (351 ca)
STRUCTURE:	Terra-Dome modular shell construction; poured concrete reinforced with steel
EARTH COVER:	2-8 ft. (.6-2.5 m)
INSULATION:	1-2 in. (2.5-5 cm) urethane foam on roof, and to 1 ft (.3 m) below frost line and exposed front of house
WATERPROOFING:	Polyurethane elastomeric rubberized membrane

HEATING DEGREE DAYS:	5,440
HEATING SYSTEM:	Passive solar, wood-burning Earth Stove™
COOLING SYSTEM:	One-ton air conditioning unit
ENERGY USE (winter 1980):	1½ cords of wood; $200 for electricity for space heat and air conditioning

Demuyt House

Located on a 5-acre (2-ha) parcel of land in Wisconsin, this house is a good example of how earth sheltering can be used to help a house work particularly well with an attractive, sloping site. As guests enter the house from the north entrance, located at a higher level than the rest of the house, they have a commanding view of a river valley. This view is duplicated at the lower-level, south/ southeast-facing front of the house.

Like most of the homes featured in this book, the Demuyt house has a primarily southern orientation in order to benefit from passive solar gain. And, as with most elevational-type homes, all the glazing in this residence is designed so that, from any area of the house, the occupants have a visual link to the outside. However, this house presents a modification of the standard elevational house design, in that

the 45-degree angles of the house permit views in three principal directions—south, southeast, and southwest.

The southeast and southwest elevations are shaded by overhangs to minimize summer heat gain; windows on the south elevation are flush with the parapet in order to maximize winter solar radiation. Deciduous trees in a planter provide shading on the south side of the house.

The interior spaces of the house are well zoned for formal entertaining and privacy of adults as well as more informal family activities. That is, the living room, dining room, and master bedroom spaces can easily be used separately from the family room and adjacent children's bedrooms. This floor plan demonstrates that a basically elevational-type house design need not be limited to a lineal

room arrangement.

The house has been designed so that an attached greenhouse could be added off the living room. In this case, passive solar gain would be achieved through the south windows of the greenhouse, and the fireplace mass would act as a wall for storing and radiating that heat.

The primary heat source for the Demuyt house is a forced-air gas furnace system, which is supplemented with heat from the fireplace and from passive solar radiation.

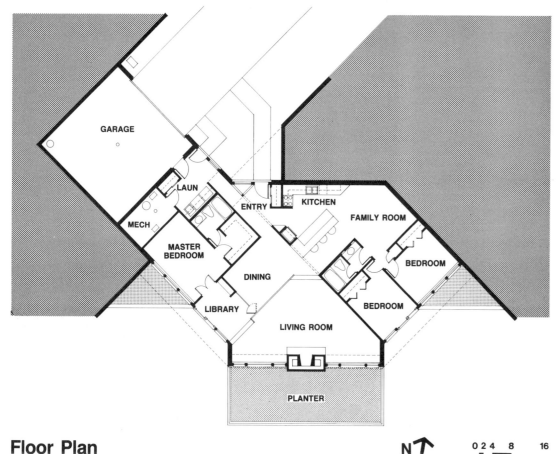

Floor Plan

GARAGE

LAUN

MECH

ENTRY

KITCHEN

FAMILY ROOM

MASTER BEDROOM

BEDROOM

DINING

BEDROOM

LIBRARY

LIVING ROOM

PLANTER

N ↑   0 2 4   8   16

LOCATION: Burlington, Wisconsin
ARCHITECT: Earth Shelter Corporation of America
CONTRACTOR: Demuyt & Son
CONSTRUCTED: Fall 1979

GROSS AREA: 2,550 sq. ft. (230 ca)
STRUCTURE: Poured-in-place reinforced concrete walls, precast concrete plank roof, concrete slab-on-grade floor
EARTH COVER: 100% on roof at 20 in. (51 cm)
48% on walls
INSULATION: Roof—4 in. (10 cm) rigid insulation
Walls—2 in. (5 cm) rigid insulation
WATERPROOFING: Bentonite on roof, Bituthene™ on walls

HEATING
DEGREE DAYS: 7,400
HEATING SYSTEM: Gas furnace, forced air system, passive solar, energy-efficient fireplace
COOLING SYSTEM: Natural ventilation

Section

LIVING ROOM

FAMILY ROOM

0 2 4   8   16

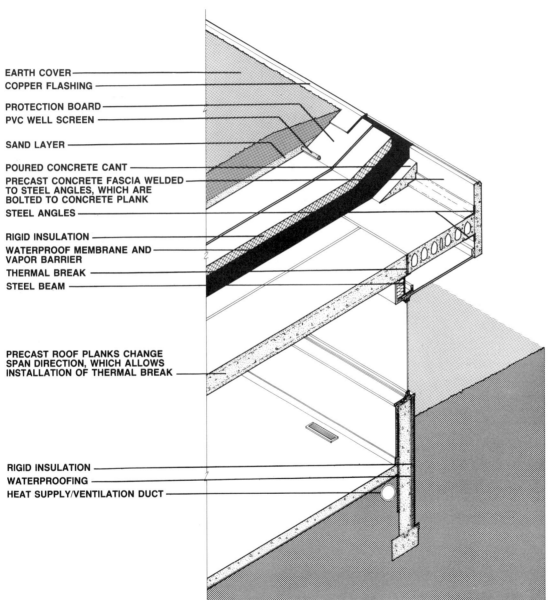

EARTH COVER

COPPER FLASHING

PROTECTION BOARD

PVC WELL SCREEN

SAND LAYER

POURED CONCRETE CANT

PRECAST CONCRETE FASCIA WELDED
TO STEEL ANGLES, WHICH ARE
BOLTED TO CONCRETE PLANK

STEEL ANGLES

RIGID INSULATION

WATERPROOF MEMBRANE AND
VAPOR BARRIER

THERMAL BREAK

STEEL BEAM

PRECAST ROOF PLANKS CHANGE
SPAN DIRECTION, WHICH ALLOWS
INSTALLATION OF THERMAL BREAK

RIGID INSULATION

WATERPROOFING

HEAT SUPPLY/VENTILATION DUCT

South Wall/Roof/Floor Section

0   1   2   4

Wheeler House

Located on an 18-acre (7.2-ha) site in rural Oklahoma, the Wheeler house uses passive solar gain in combination with extensive earth sheltering to conserve energy—a serious concern with a house this large. The earth covering also serves another important function in protecting the house from the high winds and tornadoes prevalent in this area of the country. Guests enter through a gabled entry court/greenhouse on the north; the garage, on the east end of the house, has a private entry.

This house combines an elevational-type design, featuring a south facing "garden room" area for passive solar gain, with a room arrangement similar to that of many atrium plans. A centrally located living room occupies the space where an atrium would likely be located in an atrium-type house design. However, in the Wheeler home, one double-dome skylight takes the place of the completely open roof over the atrium or courtyard. As in an atrium plan, other major living spaces—bedrooms, kitchen and dining room, and a study—are organized around the living room area (see floor plan).

On the south side of the house, the "garden room" serves as a family recreation area and painting studio for Mrs. Wheeler. The 500 square feet (46 ca) of skylights over this room make it a warm, comfortable space on sunny days, even in midwinter.

A 3-ton water-to-air heat pump provides the house with mechanical cooling and heating; passive solar gain is a secondary source of space heating. Warm air collected through the skylights in the garden room flows into the house through the upper windows between the garden and living rooms; cool air flows back to the garden room through the sliding door between the two rooms.

Another 240 square feet (22 ca) of air-type solar collectors, constructed by the architect, are connected to a rock storage bin containing a tank to preheat hot water and a copper coil to preheat water going to the heat pump.

Poured-in-place concrete waffle slab exposed on the interior forms the roof/ceiling of the house. To provide additional interior lighting, 60-watt reflector dome bulbs were mounted in boxes poured into most of the recesses of the waffle slab (see interior photo).

The Wheelers have been quite satisfied with the energy performance of their home: based on total energy cost per square foot per year, Elbert Wheeler figures that his earth sheltered home uses only half as much energy as did his previous home. Other benefits he associates with this house are the low maintenance and security afforded by earth sheltering.

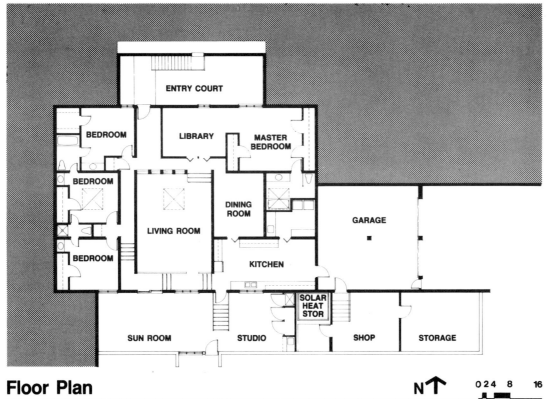

Floor Plan

BEDROOM
BEDROOM
BEDROOM
ENTRY COURT
LIBRARY
MASTER BEDROOM
LIVING ROOM
DINING ROOM
GARAGE
KITCHEN
SOLAR HEAT STOR
SUN ROOM
STUDIO
SHOP
STORAGE

N↑ 0 2 4 8   16

Section

ENTRY COURT
LIBRARY
LIVING ROOM
SUN ROOM

0 2 4   8   16

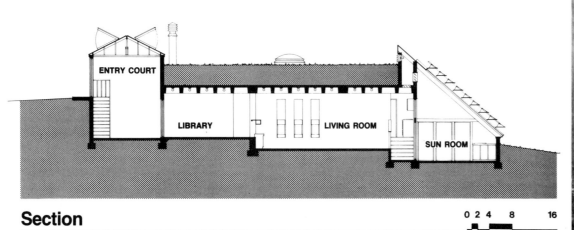

LOCATION:	Enid, Oklahoma
ARCHITECT:	Elbert M. Wheeler—Wheeler & Wheeler Architects
STRUCTURAL ENGINEER:	Louis O. Bass
CONTRACTOR:	Glen E. Payne
CONSTRUCTED:	January 1979
GROSS AREA:	5,826 sq. ft. (524 ca)
STRUCTURE:	Poured-in-place, reinforced concrete walls; cast-in-place concrete waffle slab roof; concrete slab-on-grade floor
EARTH COVER:	100% on roof at 36 in. (91 cm) 50% on walls
INSULATION:	Roof—1 in. (2.5 cm) rigid insulation Walls—1 in. (2.5 cm) rigid insulation
WATERPROOFING:	Meadows Sealtight™ premoulded membrane
HEATING DEGREE DAYS:	3,800
HEATING SYSTEM:	Water-to-air heat pump, passive solar, two wood stoves
COOLING SYSTEM:	Water-to-air heat pump

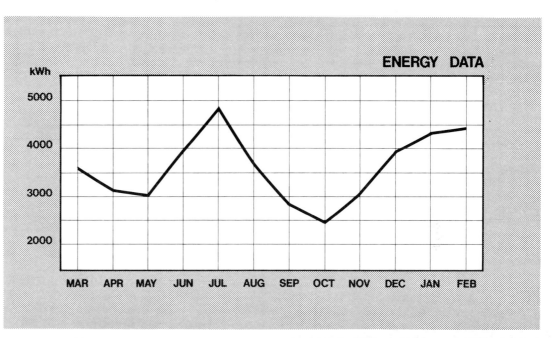

ENERGY DATA

Moreland House

The Moreland house takes advantage of the passive cooling potential of earth-integrated structures in a predominantly hot climate such as that of north central Texas. The designers say that the 3 feet (.9 m) of earth cover is quite useful in minimizing heating demands during consistently unpredictable winters.

In this case, the basic design approach involved building a wide house into a sloped hillside and installing a wall of windows across the front of the house to provide views, ventilation, and natural lighting to all major rooms. A central vented skylight provides

passive ventilation and additional lighting.

In keeping with the clients' request that their home reflect a Southwestern genre of architecture, Moreland incorporated a slanted roof and a broad porch that protects the house from the blistering sun and deflects wind and rain. The porch also allows the residents to participate in another Southwestern tradition: moving outside the house to enjoy the spring and autumn evenings. The native stone used to face the retaining walls and pave the walks also contributes to the regional character of the house.

The outward simplicity of the exterior is carried through inside the house. Rough cedar and Mexican tile give the house an inviting and relaxed quality, which is reinforced by the large brick fireplace that is the focal point of the main living space. The central skylight bathes the antique oak table with natural light.

An interesting interior design feature is the angled walls, which provide visual interest and acoustic control as both sound and light tend to diffuse throughout the rooms. The angled walls also increase the efficiency of the floor plan.

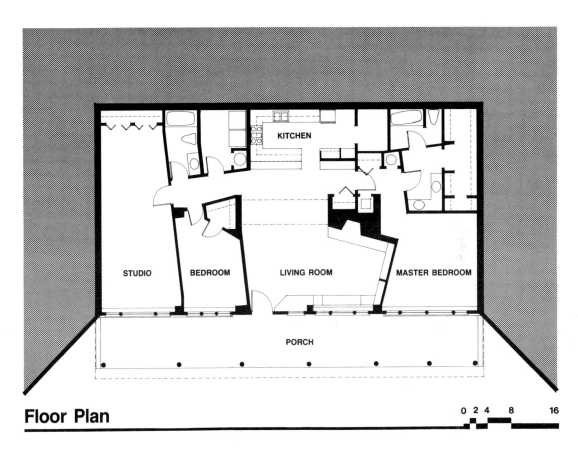

Floor Plan

0 2 4 8 16

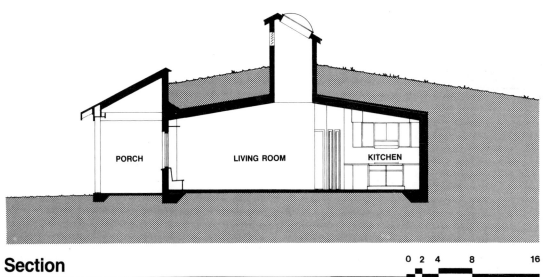

Section

0 2 4 8 16

LOCATION:	Texas
ARCHITECT:	Frank L. Moreland—Moreland Associates
CONSTRUCTED:	April 1980
PHOTOGRAPHY:	Moreland Associates
GROSS AREA:	2,100 sq. ft. (189 ca)
STRUCTURE:	Cast-in-place concrete walls; precast, prestressed concrete planks with concrete topping roof; concrete slab-on-grade floor
EARTH COVER:	100% on roof at 36 in. (91 cm) 66% on walls
INSULATION:	Roof—2 in. (5 cm) rigid insulation Walls—fiberglass batt insulation on front wall

Topic House

One of the few earth sheltered homes from which reliable, detailed energy performance data have been obtained, the Topic house is a rather typical elevational-type design. Despite dramatic northern views overlooking the Minnesota River Valley, the house was oriented south to benefit from passive solar heating and to create a more private exterior space. The house is surrounded by gently sloping hills on three sides; directly in front of the center of the house is a man-made pond.

All the major living areas have windows opening onto grade level, thus allowing a great deal of natural light to penetrate into the interior of the house and maximizing the potential for direct passive solar gain. Because the exterior surfaces have been plastered with stucco and the windows are vinyl-clad, the house is virtually maintenance-free.

During most of the winter of 1980—from January 1 through March 22—the Topic house was left unheated and was monitored for thermal performance. When the Topics left the state for the winter, they turned off all the power in the house except for the electricity needed to run one refrigerator. Temperature and relative humidity data for the house were recorded every fifteen minutes by five sensors in the house and one outside temperature sensor (see floor plan for sensor locations).

In two-and-a-half months without heat, when the lowest recorded outside temperature was –15°F (–26°C), the temperature in the body of the house never fell below 40°F (4°C). It took the house one full month for the temperature to fall from an initial 68°F (20°C) to 40°F (4°C), and for the remainder of the test period the temperature stayed within 5°F (3°C) of this figure. Even immediately inside the uninsulated windows the temperature at night never fell below 36.5°F (2°C). These data "confirmed and quantified what we thought would happen in terms of heat retaining qualities of earth sheltered homes," according to Dr. Ray Sterling,

director of the University of Minnesota's Underground Space Center.

As expected, the earth acted as a temperature moderator, so that daily temperatures inside the body of the house fluctuated very little. The temperature hardly varied at all on cloudy days; on sunny days, the maximum night-to-day temperature variation for the body of the house was only 5°F (3°C). In a conventional aboveground house under similar unheated conditions, on the other hand, temperature fluctuations would typically be considerably greater than those observed in the Topic home. The data from this house bear out experts' assumptions about the heat-retaining characteristics of earth sheltered structures.

For more information about plans for designs based on this prototype home, write to: Topic Earth Homes, 2051 Marschall Road, Shakopee, Minnesota 55379.

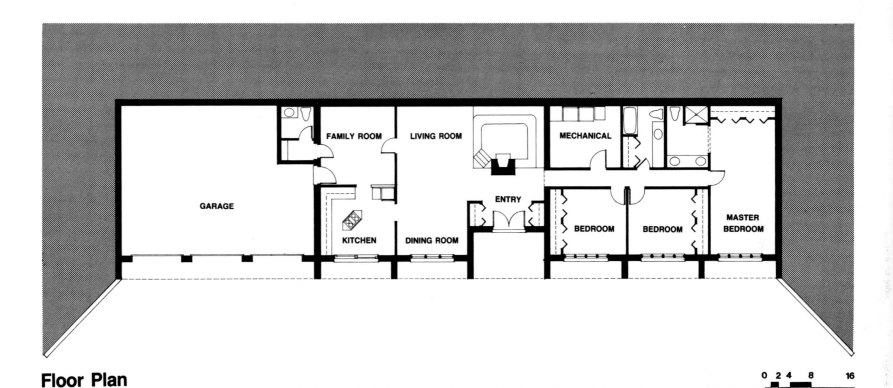

Floor Plan

0 2 4 8 16

Section

0 2 4 8 16

LOCATION:	**Shakopee, Minnesota**
CONTRACTOR:	**Joe Topic**
CONSTRUCTED:	**1977**
PHOTOGRAPHY:	**Nancy Larson**
GROSS AREA:	**2,400 sq. ft. (216 ca)**
STRUCTURE:	**Precast concrete walls, precast concrete plank roof, concrete slab-on-grade floor**
EARTH COVER:	**100% on roof at 20 in. (51 cm), sloping to 18 in. (46 cm)**
	60% on walls
INSULATION:	**Roof—4½ in. (11 cm) rigid insulation**
	Walls—2 in. (5 cm) rigid insulation
WATERPROOFING:	**Bentonize™**
HEATING DEGREE DAYS:	**8,382**
HEATING SYSTEM:	**Heat pump with electric furnace**
COOLING SYSTEM:	**Heat pump/air conditioner**

Topic House Data — Winter 1980

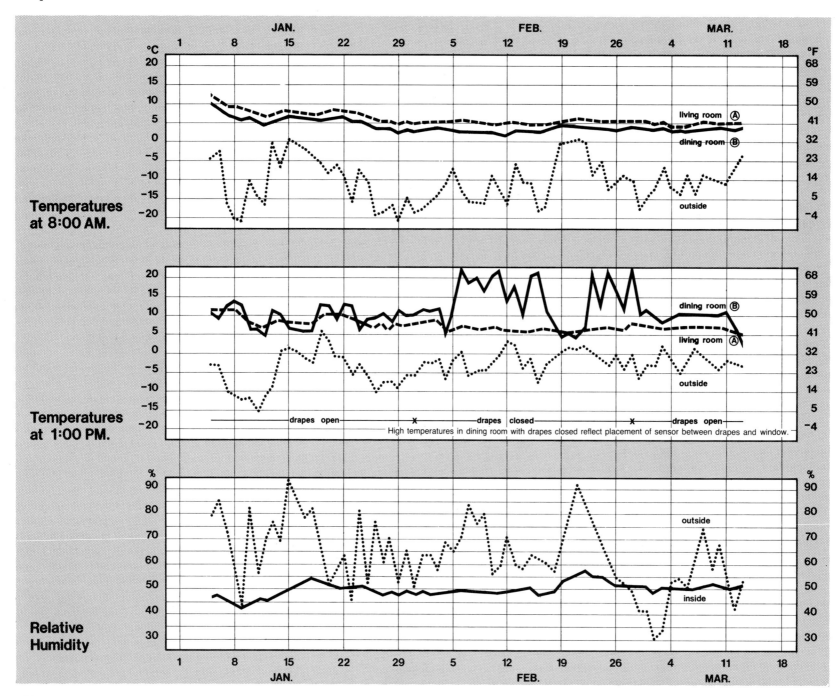

Temperatures at 8:00 AM.

Temperatures at 1:00 PM.

drapes open —— X —— drapes closed —— X —— drapes open

High temperatures in dining room with drapes closed reflect placement of sensor between drapes and window.

Relative Humidity

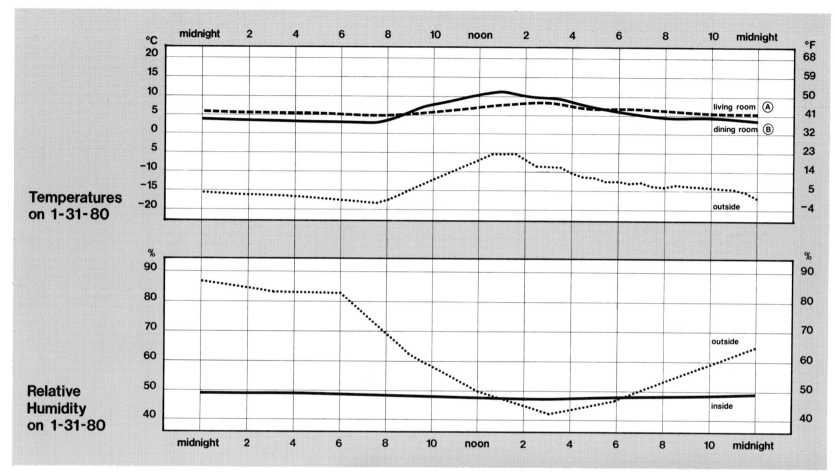

Temperatures on 1-31-80

°C: 20, 15, 10, 5, 0, 5, −10, −15, −20

°F: 68, 59, 50, 41, 32, 23, 14, 5, −4

midnight, 2, 4, 6, 8, 10, noon, 2, 4, 6, 8, 10, midnight

living room (A)
dining room (B)
outside

Relative Humidity on 1-31-80

%: 90, 80, 70, 60, 50, 40

outside
inside

midnight, 2, 4, 6, 8, 10, noon, 2, 4, 6, 8, 10, midnight

Appendix

We wish to thank the following designers and builders of the houses featured in *Earth Sheltered Homes*:

Architerra, Inc.
Rosslyn Center
1700 North Moore Street
Arlington, Virginia 22209

The Architectural Alliance
400 Clifton Avenue South
Minneapolis, Minnesota 55403

Boothe & Associates
5608 Malvey
Fort Worth, Texas 76107

Norm Clark, Architect
6730 S.W. Parkwest
Portland, Oregon 97225

Close Associates, Inc.
3101 East Franklin Avenue
Minneapolis, Minnesota 55406

Colorado Sunworks
Post Office Box 455
Boulder, Colorado 80306

Design Consortium, Inc.
1012 Marquette Avenue
Minneapolis, Minnesota 55403

Earth Shelter Corporation of America
Route 2 Box 97B
Berlin, Wisconsin 54923

Earthtech
Post Office Box 52
Lyme, New Hampshire 03768

Ellison Design and Construction Company
2001 University Avenue
Minneapolis, Minnesota 55414

Richard Engan Associates: Architects
323 4th Street West
Post Office Box 89
Willmar, Minnesota 56201

McGuire/Engler/Davis/Architects
423 South Main Street
Stillwater, Minnesota 55082

Moreland Associates
904 Boland
Fort Worth, Texas 76107

SEAgroup
418 Broad Street
Nevada City, California 95959

Sticks & Stones Design
2201 21st Avenue South
Minneapolis, Minnesota 55404

Terra-Dome, Inc.
14 Oak Hill Cluster
Indepedence, Missouri 64057

Joe Topic
2051 Marschall Road
Shakopee, Minnesota 55379

Richard F. Webster, A.I.A.
663 Merrimon Avenue
Asheville, North Carolina 28804

Malcolm Wells, Architect
Post Office Box 1149
Brewster, Massachusetts 02631

Wheeler & Wheeler—Architects
119 North Washington
Enid, Oklahoma 73701

Index

Active solar technology ... 11-12
Architerra houses ... 53, 96-101
Atrium house designs ... 13, 15
Barnard, John ... 11, 16
Basement houses ... 16
Boothe house ... 88-91
Burnsville house ... 9, 14, 21, 34-37
Camden State Park house ... 21, 22-25
Clark house ... 13, 53, 62-65
Demuyt house ... 106-109
Earth Shelter Digest ... 8
Earth Sheltered Community Design ... 17
Earth Sheltered Housing Design ... 8, 17
Earthtech houses ... 13, 53, 78-83
Ecology house ... 16
Elevational house designs ... 12-13
Feuille house ... 12, 53, 70-73
Geier house ... 13, 14

Hadley house ... 14, 53, 66-69
History of earth sheltered
  construction ... 15-17
Johnson, Philip ... 16
Landscaping ... 14
Metz, Don ... 16
MHFA Demonstration Project ... 17, 21
Moreland house ... 53, 114-117
Oklahoma State University ... 17
Passive solar technology ... 11, 21, 53
Remington house ... 14, 53, 58-61
Rolla, Missouri house ... 11
Seward town houses ... 9, 12, 21, 38-41, 96
Sod houses ... 15
Soil temperature fluctuations ... 10
Solar Energy and Energy Conservation
  Bank Bill ... 17

Sticks & Stones house ... 53, 62, 74-77
Suncave ... 53, 54-57, 58
SunEarth house ... 53, 84-87
Terra-Dome house ... 53, 102-105
Texas Tech University ... 17
Topic house ... 11, 118-121
Underground Space ... 8
Underground Space Center ... 8, 17, 18, 118
University of Missouri-Rolla ... 17
University of Texas-Arlington ... 17
Waseca house ... 21, 30-33
Wells house/office ... 92-95
Wells, Malcolm ... 16, 92
Wheeler house ... 53, 110-113
Whitewater State Park house ... 12, 21, 22, 46-49
Willmar house ... 21, 26-29
Winston house ... 14, 16